The Practical Geologist

How To Apply Primitive Skills For Everyday Use

GARRET ROMAINE

FALCON

ESSEX, CONNECTICUT

To my nieces Sydney and Taylor—a pair of practical super Moms

An imprint of Globe Pequot, the trade division of
The Rowman & Littlefield Publishing Group, Inc.
4501 Forbes Blvd., Ste. 200
Lanham, MD 20706
www.rowman.com
Falcon and FalconGuides are registered trademarks and Make Adventure Your Story is a trademark of
The Rowman & Littlefield Publishing Group, Inc.

Distributed by NATIONAL BOOK NETWORK

British Library Cataloguing in Publication Information available
Library of Congress Cataloging-in-Publication Data available

ISBN 978-1-4930-6213-3 (paper: alk. paper)
ISBN 978-1-4930-6214-0 (electronic)

∞™ The paper used in this publication meets the minimum requirements of American National Standard for Information Sciences—Permanence of Paper for Printed Library Materials, ANSI/NISO Z39.48-1992.

CONTENTS

PREFACE

This guide is aimed at everyone who is interested in learning enough about geology, topography, soil science, and rocks and minerals to put that knowledge to use. We'll explore the many ways to turn common materials into resources in order to make improvements, survive the elements, or otherwise gain an edge. We'll study the ancients and learn how they adapted and overcame the world around them to not just survive but thrive.

A trip outdoors is like exploring nature's hardware store; the better you understand enough geology to be dangerous, the faster you can put that knowledge to work. For survivalists, campers, hikers, or primitive skills aficionados, geology can mean the difference between life and death. So we'll start with the ways geology can augment survival skills in the search for water, shelter, and food. Then we'll move on to the concept of homesteading and show how geology skills can make all the difference.

In this book, we'll assume that you're an interested newcomer to the language of geology and try to explain things as we go along. It may seem a bit bewildering at first, because geology borrows from chemistry, physics, math, chemistry, biology, and more to create its own language. Fortunately, many of the best metaphors come from cooking. In a sense, the Earth is like a big kitchen, constantly mixing and remixing ingredients, skimming some off, concentrating others, and recycling everything in a constant churn. We'll end up using a lot of cooking metaphors and examples, trying to make the science as practical as possible.

Our ancestors faced enormous hurdles in getting enough to eat and overcoming the challenges of storing, preparing, hunting, and gathering food. Still, they had several advantages over their competitors: opposable thumbs to fashion tools; a powerful, curious brain to puzzle out complex situations; and a sense of community with powerful bonds forging strong personal relationships. In many ways, survivalists retrace those steps when they hone their skills. If you can learn to tap into the connection our ancestors had with the land around them, you'll be comfortable searching out the best rocks for fashioning into tools and weapons. You'll learn the best materials for building shelters, and you'll draw on their knowledge for insights into what rocks will come in handy as you make improvements. If you can appreciate the skills required to survive advancing glaciers, continual drought cycles, daunting floods, and other geo-catastrophes, you can understand how early humans were the first practical geologists, and well worth studying.

There are many procedures and recipes sprinkled throughout this book, but there is no substitute for your own trial and error. Your mileage may vary on just about anything related to geology—your limestone may not be pure enough to convert to slaked lime, or your soil may have too much sand to turn into cob for housing. Your sand may be too round, or your water too full of calcium. The science is sound, but the ingredients you have at hand may require adjustments to the standard procedures, ratios, and recipes. Fortunately, there are even more resources available if you want to dive deep into a certain activity. There are countless YouTube videos, how-to books, detailed guides, academic studies, and more to use for additional research. This guide is meant to give you ideas, show you what's possible, and inform you about how the ancients solved similar problems. Have fun, be patient, and keep at it—you never know when these skills may be absolutely vital once again.

ACKNOWLEDGMENTS

Many thanks to the staff at the Rice Museum of Rocks and Minerals in Hillsboro, Oregon, with special shout-outs to executive director Aurore Giguet, curator Angela Piller, office manager Lena Toney, and store manager Dylan Roost. They opened up the collection for photographs, shared insights on basic geology, and made life easier during the inevitable deadline crunch.

Let me also express additional gratitude to mineralogist Julian Gray, who provided assistance with explanations and patience with the manuscript. Field workers Nick Hensen and Frank Higgins helped with several trips into the wilds to track down survival tips, and spent a couple nights inside a pair of Nevada mine adits to test out assumptions. Survivalist Jake Riley provided a few tips and hacks, as did Dirk Williams, on multiple camping expeditions.

And finally, thanks to my wife, Cindy, for her patience through yet another project schedule, and her willingness to ignore a workbench full of specimens, long-running experiments, multiple open books on the floor, and one more trip to the store. Especially, thanks for her grudging acceptance of the need to upgrade the trusty old Jeep to the Rubicon class for the ultimate "bug-out" vehicle.

Good field test for agate—shine a flashlight under it. Jasper, chalcedony, quartzite, and plain quartz don't blaze like this pleasing chunk.

INTRODUCTION

Our ancient ancestors were rockhounds. They read the rocks around them and created tools, art supplies, weapons, fireplaces and hearths, housing, and homesteads. While practical geology was only one facet of their encyclopedic knowledge of the world around them, it was an important ingredient. They combined geology, chemistry, biology, botany, physics, and more to first survive, and then thrive.

That's the model this book follows, starting with survival tips based on geology, then focusing on how civilizations rose from the Stone Age to where we are today. There's a basic primer on geology to give you enough information to get started, but some of you may just skim that section, and that's fine.

We'll end up discussing how practical geology is guiding our growing space exploration efforts. As humans investigate the solar system, many of the same skills and instincts are coming back into play.

How It Started

In this book, you'll learn how primitive cultures constantly took stock of the resources around them, and why they're a good model to study for insights today. The ancients identified and exploited salt licks, ochre deposits, chunks of pyrite, and scattered obsidian beds with great efficiency. They explored caves, erected cairns along trails as navigation aids, and used rocks to grind seeds. They were the original practical geologists.

We'll investigate advances from the Stone Age to the Copper Age, then the Bronze Age, and finally the Iron Age, with an eye on how ancient metallurgists progressed as furnaces, forges, smelters, and kilns grew in sophistication. We'll look at ancient construction methods with adobe bricks, clay bricks, cob, rammed earth, and finally the perfected recipe for Portland cement. Many of the topics and discussion points will prove useful to survivalists, campers, and trekkers. In essence, this book can help convert a homestead into a hacienda by studying the geology and utilizing it fully.

For example, consider how primitive cultures used rock dams across small creeks to trap fish, which they could then scoop out of the water with bare hands, if necessary. That's some practical geology for sure. A horseshoe-shaped rock fish trap works similarly in a larger river, as the fish seek refuge against the current and hide in the calm. Hunter-gatherers might move beyond bare hands and use gaffs, spears, nets, baskets, or other tools to complete the task. Those fishing tools would have required sharpening, shaping, cutting, smoothing, and other manipulation using stone tools such as knives, borers, scrapers, cutting edges, hammer stones, hand axes, and more. It really did usually come down to geology to survive and thrive.

Constant and Important Breaking News

These are exciting times for those studying ancient life. Discoveries are taking place at a dizzying pace, and thanks to the internet, we get the news quickly, and with context. Websites such as ScienceDaily, Jerusalem Post, Indianculture.gov, the Journal of Islamic Archaeology, and more are quickly peeling back the pages of time all over the world. It appears not a week goes by without some announcement of interest, from retreating glaciers in Europe revealing ancient artifacts to human fossil discoveries in Africa filling in the gaps of our family tree.

As just one example, right as this manuscript was in final markup, researchers writing in the journal *Antiquity* announced the discovery of 2,000-year-old cave art in Alabama. Giant glyphs carved into the mud surface of the cave ceiling, which depict human forms and animals, are some of the largest known cave images found in North America. Using a technique known as reflectance transformation imaging (RTI), which is similar to photogrammetry, 3D models revealed complete and complex images that might have gone unnoticed just a couple decades ago.

A key benefit is that long-held ideas about gender roles in ancient societies are getting a much-needed update. For example, the historians behind *Lady Sapiens: The Woman in Prehistory,* a documentary to be published in 2022, aim to debunk the simplistic division of ancient gender roles. Using advances in the study of bones, graves, art, and ethnography often ignored in the public sphere, these researchers want to banish the idea of women as helpless, frightened creatures, protected by overly powerful male hunters.

Excavating burial sites from 10,000 to 14,000 years ago in North America, researchers noted that skeletal evidence showed well-developed upper arms in women, and many were buried with weapons and tools. One Peruvian tomb of a young woman contained twenty-four stone artefacts, comprising a toolkit of everything needed to hunt and butcher big game: six projectile points, four scrapers, a knife, and several chipped flakes of stone.

Clearly, both ancient men *and* women were practical geologists. And currently, with technological barriers breaking down across scientific disciplines, and with humankind stretching beyond Earth for distant worlds, we must follow the science and see where it leads us. That's what this book attempts to do—to make the science of geology more accessible and easier to understand, to show how it impacts us to this day, and perhaps, to see where it might lead us.

Fine chunk of obsidian from Oregon's famed Glass Butte, showing rare rainbow colors and a nice conchoidal fracture.

CHAPTER 1
GETTING STARTED

On July 21, 1914, an intrepid survivalist named Joseph Knowles said goodbye to a group of journalists, stripped off all his clothes, and plunged naked into the Siskiyou Mountains of southern Oregon. He was starting a sixty-day "test" to prove that a modern man was still enough of a sturdy barbarian to survive on his wits alone, with no clothes, no tools, and no assistance. Knowles had conducted the same "experiment" a year earlier in the woods of Maine, with similar publicity.

In his Maine foray, Knowles was tapping into a rising fear in the early 1900s—that modern humans were becoming soft. While complex machinery pumped out an increasing amount of time-saving and labor-reducing devices, assorted essayists and pundits fretted that we humans were losing our survival instincts. President Theodore Roosevelt, who had grown up as a sickly, pampered child but overcame that start by embracing the outdoors, wrote in 1899: "Unless we keep the barbarian virtues, gaining the civilized ones will be of little avail."

From his primitive shelter in Maine, Knowles echoed that concern in a letter to President Woodrow Wilson in 1913, purportedly scrawled on birch bark using charcoal-tipped sticks. "My object is to demonstrate that modern man is not only the equal of primitive man in ability to maintain himself, but that civilization has so improved the human mind that he may add to primitive life accomplishments which our early ancestors never knew," Knowles wrote.

After battling the New England elements to a draw, Knowles declared victory and returned to civilization. There was grumbling in the local press about some of the details, with suspicions about his getting assistance. For his Oregon adventure in an even more remote wilderness, he would repeat the "man vs. nature" ordeal, and this time leave no doubt.

Knowles soon ran into trouble—rampant poison oak reduced his feet to a blistered, infected mess, and the cold climate at higher elevations pushed him lower down, into the valleys. He lived on roots, berries, and nuts, augmenting that diet with fish, which he speared. He flaked agates into slicing tools for cutting bark and vines, which he wove into sandals and a loincloth. Eventually he came across the carcass of a deer and set about tanning the hide using a variety of stone tools, including a hammer stone and a scraper. He used a friction method to get a fire going and built a debris hut with stones, sticks, and moss for sleeping.

Within thirty days, he again cut things short and called the stunt a success, emerging from the wilderness fully clothed in tanned hides and sturdy footwear. But instead of the nationwide press and hordes of well-wishers he expected, events in Europe demanded all the headlines. Archduke Ferdinand had been assassinated in Sarajevo, precipitating World War I. Knowles paraded around the streets of Grants Pass, Oregon, in his handmade clothing, and a local brass band serenaded him, but the world's attention was elsewhere. He would attempt one more foray, this time with a female counterpart dubbed "Dawn Woman," in the Adirondacks, but the Broadway starlet they selected gave up after a week, covered in bug bites.

In the twenty-first century, we still worry about our supposed softness as life goes on. We fear the real possibility that all this modern convenience can be snatched away in an instant. From asteroid impacts to a zombie apocalypse, we know we might instantly face the same hardships as our ancestors—seeking food, needing a fire, and requiring a shelter in a hurry. Indeed, many successful television programs push the idea that survival may hinge on our wits and attitudes, if not our ingenuity and mental fortitude.

Naked and Afraid (Discovery) reprises the old "Dawn Woman" stunt dreamt up by Joseph Knowles with a modern twist. Contestants go into the wild with no clothes but can choose one tool, frequently bringing a machete or fire starter, and are often supplied with a cooking pot to boil water. The human drama is intense; personality clashes are a mainstay. Producers place the hapless humans in a wide variety of harsh environments, and frequently the smallest wildlife—gnats, ticks, mosquitoes, and bacteria—are the real winners.

Dual Survivor (Discovery) starred two survivalists, always male, from different viewpoints to battle the elements. Cody Lundeen, a primitive skills expert from the Arizona desert, was the perfect foil for the macho warrior creed of those he was often teamed up with. He highlighted his laid-back approach by walking barefoot, which he said forced him to slow down and pay more attention.

Cody was eventually replaced by Matt Graham, who also studied aboriginals and adopted their methods, but did wear sport footwear. In season 5, episode 1, "Into the Canyons," Matt and Joe Teti (a former

A rockhound would see agate here, but a survivalist sees resources for stone tools.

soldier) were put in the position of a mountain biker who got too far out in the Utah desert and broke down. Teti's military training made him inclined to strip the bike of anything that might be a resource and haul the booty off, but Graham talked him out of it. He told Teti that they'd find everything they needed, that they didn't have the time or energy to waste on tearing down and hauling out the machine parts.

Graham turned out to be right. He guided them to a little green valley with running water where he fished with his hands and located a big chunk of what looked like seam agate, which he proceeded to break down into shards. Soon they had fashioned stone knives, started cutting and weaving cordage, built a friction fire, and begun to not just survive, but thrive. All with the resources they found around them, and starting with the most basic stone tools.

Man vs. Wild (Discovery) featured the intrepid Bear Grylls, a former British special forces warrior. With boundless energy and an iron-lined stomach, Grylls tackled deserts, snowfields, mountains, and forests with equal enthusiasm. He fashioned ropes from vines and other materials to scale cliffs and canyons; built shelters out of rocks, trees, bushes, and scavenged cordage; and was fearless exploring caves. He used clay-rich mud to protect against sunburn; fashioned water filters out of pebbles, sand, and charcoal; and once converted a dead seal carcass into an upper-body wet suit to protect his most vital organs for a brisk swim. He enjoyed a good hot springs soak in Iceland and inspected an abandoned mine in the Dakotas for resources. He expanded the show concept eventually to take athletes, movie stars, and other celebrities on overnight junkets, inviting them to share his diet of bugs, worms, snails, and whatever else he could forage.

Primal Survivor (National Geographic) features another survival expert attuned to the ancient ways. Wilderness guide and survival instructor Hazen Audel intensely studies aboriginal tribes and native skills experts and uses them in his show. He's able to demonstrate age-old survival techniques, many of which rely on a practical knowledge of geology. On the show he harnessed fire to build a dugout canoe, used a blowgun with poison-tipped darts, wove baskets from leaves and reeds, fashioned arrowheads from local rocks, and demonstrated key survival knowledge handed down through countless generations.

In addition to the ideal of modern humans successfully reverting to their hunter-gatherer roots, there is also a growing homesteading movement viewable on TV. Another popular Discovery series, *Homestead Rescue*, features a family of successful Alaska homesteaders who visit failing or near-failing homesteads across North America and teach them tips, tricks, hacks, and shortcuts to turn things around. Patriarch Marty Rainey and his clever kids are masters of geology hacks—building temperature-controlled root cellars, fashioning goat pens out of caves, digging diversion canals to control floods, and using mud and straw "cob" to finish the exterior of a building made with old tires. They even showed homesteaders how to recover placer gold for a revenue source. Their use of practical geology skills dominates their homestead hacks.

In just about every scenario in these popular shows, a basic understanding of geology comes in handy. Whether creating hand tools on the fly, seeking out a cave for shelter, filtering water, restoring soil, or building simple structures, these shows illustrate how important it is to view our environment as a complete hardware store. Just having a basic understanding of rocks and minerals can mean the difference between drinking clean water and dying from amoebic dysentery.

So that's where this book comes in. We'll begin with basic survival scenarios and work our way through them. We'll explore the way humans began taking advantage of natural resources long ago and carry

Bear Grylls probably wouldn't think twice about tying off a rope and checking out this old mine, but you should stay out.

that through to modern times. You'll see how geology can shape a successful homestead as well as a weekend outing. We'll cover some of the many hazards our restless Earth can throw at us, and what you can do to avoid them. And if you ever do find yourself in an emergency, many of the common-sense techniques discussed here will come in handy.

CHAPTER 2

UNDERSTANDING GEOLOGY

"Practical geology" can be described as the bits and pieces of a large body of earth sciences that actually apply to everyday life. You'll benefit if you're interested in collecting rocks and minerals, doing a little prospecting, having more fun camping, surviving in the wilderness, or starting or expanding a homestead. We will start with general earth processes, mostly to give you insights into where to go and what to look for.

Geology is a young science, dating back only to the early 1800s as a formal field of study. But the ancients knew how to use science to become better miners, studying economic ore deposits and developing theories about earth catastrophes. They used geology hacks for building structures, transporting water and waste, and improving agriculture. But before explaining the many ways humans have learned to adapt and thrive, it will be useful to go over some key concepts.

ALL ABOUT MINERALS

Minerals are the building blocks for rocks. You should understand some very basic concepts about chemistry in order to you help identify minerals, so they're listed here. You can make an entire career out of chemical engineering and materials science, but for now, let's cover the essentials and talk about the things you need to know as a rockhound to perform bush fixes and geology hacks later on.

Our Chemical Building Blocks

Chemists have worked out a simple method for organizing the elements, which are the foundations of every rock and mineral. Think of the hierarchy like this:

The earth is composed of	Core, mantle, and crust
The crust is composed of	Tectonic plates
Plates are composed of	Rocks
Rocks are composed of	Minerals
Minerals are composed of	Atoms, or elements

Every practical geologist should know a little about the elements in the periodic table, shown below.

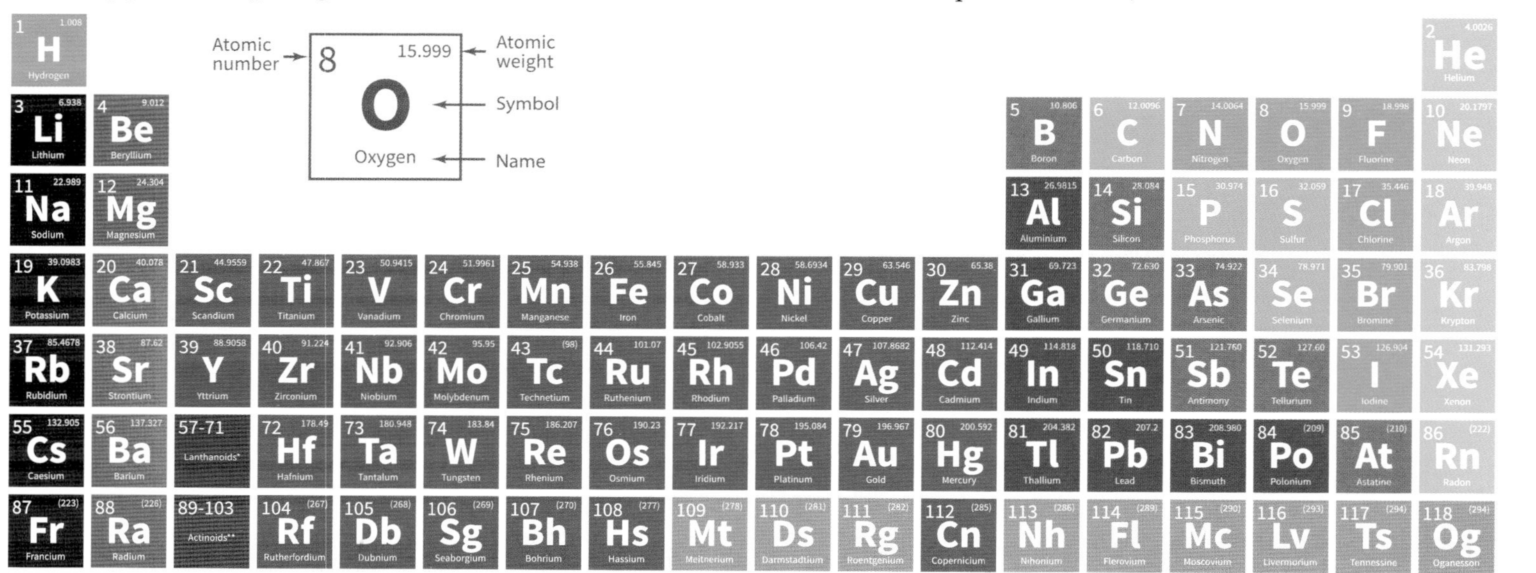

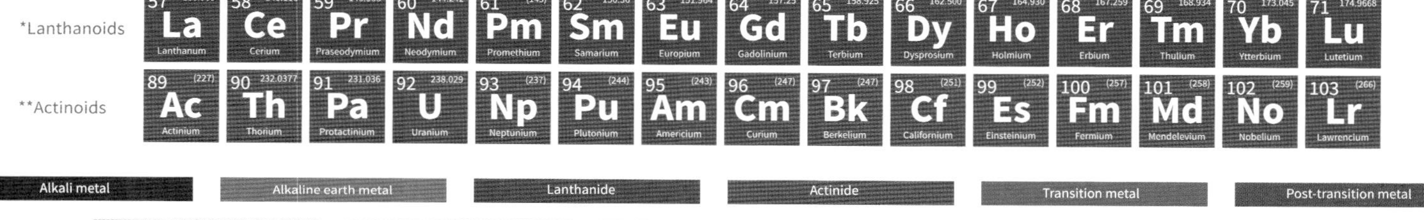

Periodic table listing all known elements ALEXLMX/GETTY IMAGES

The Periodic Table

Humans have long struggled to make sense of the rocks and minerals around them. In the fifth century BCE (before the Common Era, or the year 0), Leucippus and his pupil Democritus proposed that all matter was composed of small indivisible particles. They called these particles atoms, from the Greek *atomon*, for "uncuttable," or "indivisible." Democritus proposed that nature fundamentally behaves as a machine or a highly complex mechanism—an early expression of the Gaia concept.

Metal elements were a key early interest—gold and silver had applications in jewelry, while copper, bronze, and iron made successively harder tools and weapons. Copper is widely assumed to be the first metal humans mastered. The reason for this is simple: Native copper exists in elemental form as copper nuggets and lumps in Middle East nations such as Iran. Early metalsmiths soon learned that if you heated a metal, you could perform any number of useful actions on it. The scientific term is annealing—a process that softens and improves the ductility (and/or toughness) of copper and copper alloys. The process involves heating, holding (soaking), and cooling. Copper has a relatively low annealing point; at about 400°C (752°F) you can start to improve its qualities.

Ancient metal workers learned how to smash a copper nugget like this one into tools before they figured out they could melt the metal and pour it into a mold.

A team of British, Serbian, and German archaeologists working under a grant through the University College London (UCL) showed that the most likely center of early metallurgy was in the highlands of modern-day Iran. In the 2008 paper *New Insights into the Ancient Mining and Metallurgical Researches in Iran*, the authors indicate that the earliest examples of crude copper metalsmithing, simple hammering and annealing, are reported from west of the Zagros Mountains in northeastern Iraq about the late tenth millennium BCE. In eastern Anatolia, simple copper tools date to the late ninth millennium BCE.

Once early workers began experimenting with copper nuggets, they probably soon figured out they would need to improve their fire pits and began building crude smelters to achieve higher temperatures. One key requirement to make a hotter fire is a bellows to move more air; another key necessity is plenty of charcoal. Archaeological excavations show that by 9000 BCE, copper working had reached highly technical levels in Iran, Mesopotamia, and Egypt.

Lead, another soft and easily smelted metal ore, was probably next. At the ancient Hittite city of Çatalhöyük (Catal Huyuk) in Turkey, beads of lead have been recovered that date back to about 6500 BCE. Gold, silver, iron, tin, mercury, and zinc soon followed, as metallurgists continued unraveling the various elements.

This specimen of galena (left), from the Lead Belt of southeast Missouri, shows the characteristic square angles of good ore. Lead weights such as the one on the right, for balancing tires, are a common find on most US streets.

Table 1: Discovery of Elements in Rough Order

Element	Symbol	Estimated Discovery	Location
Copper	Cu	9000 BCE	Turkey, Iran, Iraq
Lead	Pb	7000 BCE	Africa
Gold	Au	6000 BCE	Israel
Silver	Ag	5000 BCE	Asia Minor
Iron	Fe	5000 BCE	Egypt
Carbon	C	3750 BCE	Egypt, Sumer
Tin	Sn	3500 BCE	Turkey
Sulfur	S	2000 BCE	Egypt
Mercury	Hg	1500 BCE	Egypt
Zinc	Zn	1000 BCE	India

Ancient chemists had discovered arsenic, bismuth, and antimony by 1000 BCE. Another fruitful round of discoveries began in 1669 with the discovery of phosphorous, recovered from urine by German alchemist Hennig Brand. Then, in 1735, a succession of elements emerged: cobalt, platinum, nickel, magnesium, hydrogen, and oxygen. By 1787, nine more new elements had been identified, marking thirty-three known elements. It was time to get the entire family documented in an easy table to sort them out.

The American Society for Biochemistry and Molecular Biology (asbmb.org) has a solid discussion about the origins of what we now call the periodic table. The paragraphs below are excerpted from that 2021 article by Deboleena M. Guharay.

In 1789, French chemist Antoine Lavoisier (1743–1794) attempted an early chart that grouped all known elements as either metals or nonmetals. Sometimes known as "The Father of Modern Chemistry," Lavoisier was stymied by several obstacles. He couldn't list the elements in order of increasing atomic number (the number of protons in the nucleus of an atom) because the idea of atoms being made up of protons, neutrons, and electrons was still unknown. He managed to sort elements into gases, nonmetals, metals, and earths.

In 1829, German physicist Johann Wolfgang Döbereiner (1780–1849) observed an interesting pattern among certain elements. He found that the properties of bromine, a liquid, were halfway between those of chlorine gas and the solid iodine. He identified comparable combinations and relationships in two other "triads"—calcium, strontium, and barium; and sulfur, selenium, and tellurium. The idea that chemists could anticipate "missing" elements soon caught on.

French chemist Jean Baptiste André Dumas (1800–1884) eventually perfected a method for measuring vapor densities and thereby determining atomic weights. In an 1826 paper, Dumas described measuring thirty of the elements. At an 1860 conference the attendees agreed that hydrogen would be assigned the atomic weight of 1 and the atomic weight of other elements would be decided by comparison with hydrogen.

By 1863, there were fifty-six known elements, with a new element being discovered at a rate of approximately one per year. British chemist John Newlands (1837–1898) arranged the elements into a periodic table with increasing order of atomic masses in 1865, when he showed that every eight elements had similar properties and called this the Law of Octaves.

Russian chemist Dmitri Mendeleev (1834–1907) developed the periodic table as we know it today in 1869. Mendeleev used the previous work of multiple contributors, but he made several breakthroughs. He placed the elements into columns, which he called groups, and into rows, which he called periods. He noted the blocks of elements with similar properties, such as whether they are a solid, liquid, or gas at room temperature and their ability to bond with other atoms. His work, while incomplete, still stands as an important scientific milestone.

Simple Crystals

You probably remember from your science classes that ions are atoms with an electrical charge, either positive or negative. Water, or dihydrogen monoxide, is two (di) hydrogen atoms and one (mono) oxygen atom—H_2O. Individual atoms tend to have positive or negative charges, and since opposites attract, that's how molecules are formed. The math is simple: A single oxygen ion typically has a strong, negative 2 (-2) electrical charge. Each hydrogen has positive 1 (+1) electrical charge. Molecules are only stable when the math balances out, so:

$$[2 \times (+1)] + [1 \times (-2)] = (+2) + (-2) = 0$$

Salt, or sodium chloride, is even simpler: NaCl. One sodium (Na, for the Latin *natrium*) atom with a positive charge combines with one chlorine (Cl) atom with a negative charge to create what geologists refer to as halite, or common salt.

Gold, symbolized by the initials Au, for the Latin *aurum*, doesn't react with many other atoms. You won't find gold oxide or gold chloride; the best you'll do is find gold with tellurium, forming a telluride, or locked up in an arsenic-rich pyrite called arsenopyrite. Gold is most often found as an element (with silver impurities); thus its chemical formula is just Au. If there is enough silver to color the gold a light gray, the ancients coined the term "electrum" for the result.

Gold nuggets and flakes are rarely pure gold; they usually contain varying amounts of silver.

Some other elements that are found in their native form are silver (Ag, for the Latin *argentum*), platinum (from the Spanish *platina*, "little silver") with the atomic symbol Pt, copper (Cu, for the Latin *cuprum*), and sulfur, shown as a simple S. Diamonds are pure carbon (C); noble gasses such as neon (Ne), argon (Ar), and Krypton (Kr) are also pure elements. However, these are all the exception—most elements happily react with another element, such as oxygen (to form oxides) or sulfur (to form sulfides), or oxygen *and* sulfur (to form sulfates). That's where things get more complicated. First, let's create a simple crystal from something you find easily in most well-stocked kitchens.

There are several simple crystal-growing experiments you can perform in the kitchen with your kids. Do an internet search for "creating salt crystals" or "creating alum crystals" and you'll find all kinds of easy labs. You'll find step-by-step instructions for how to grow crystals from sugar, salt, and alum. The basic idea is to supersaturate heated water with as much material as possible and then evaporate the water, leaving the crystals behind to begin building around a seed crystal.

The concept of a seed crystal is important. As the solution cools and loses energy, the molecules start looking for something to form around; but if everything nearby is molten, there is nothing to begin connecting to. In a vein or vug, where you have a small chamber full of chemicals in gas or liquid form, the seed crystals may form along the walls of the chamber.

Rockhounds only need to understand a few more basics about chemistry and how atoms combine. Those molecules can line up in repeating combinations that form a crystal lattice, and there are several common forms for crystal shapes, including cubic, as in the salt example, or hexagonal, as for quartz and beryl. These common shapes are called crystal habits, and there are a lot of them—needles, blades, plates, and many more. Consult the "References" section or find some books at your local library if you want to learn more; mineralogists and crystallographers can spend careers looking for new minerals or understanding old ones. Some of the terms are real tongue twisters, such as "botryoidal" (pronounced bah-TROY-dal). It refers to the round, bubbly structure you sometimes see with jade or malachite. Rockhounds usually just say the specimen has "bots" and move on.

We actually don't see a lot of obvious minerals out in the field; we mostly see rocks with microscopic minerals, which we'll talk more about next. The best places in the field to look for minerals are the tailings piles at mines and in quarries. Some of those places can be dangerous; road cuts and cliffs also require caution. Tailings piles, where miners piled up waste rock or marginal ores, offer the best opportunity, but always use caution around old mines.

The oxygen in our atmosphere starts to attack fresh minerals quickly and begins to oxidize them if possible. You've seen rust on fresh steel, so you know about oxidation. The science for creating rust, which is iron oxide (Fe_2O_3), is straightforward. Basically, you take four iron atoms (Fe), add three oxygen molecules (O_2), and you get two iron oxide molecules.

Many economic ore deposits are associated with pyrite, or FeS_2. Pyrite is the most common sulfide mineral, and is often a clue that there was something good occurring in a quartz vein. Pyrite is very susceptible to oxidation as water separates the iron from the sulfur, leaving red and yellow streaks. Prospectors grew to appreciate rusty staining in a quartz vein. Such staining told prospectors that there was probably some interesting mineralization if they could get a fresher sample from the vein, where water had not already started attacking. If gold or silver was present in an oxidized ore body, the ore had probably freed up the gold, making it easier to process. Many a boomtown in the western United States started as miners exploited the easy, oxidized zone near the surface. Once the fast riches were gone, a deposit might turn out to be marginal at best, and the town would fold as miners fanned out and searched for another boom.

Quartz with rusty staining is a telltale clue for prospectors.

Mineral Identification

In this section we'll discuss some of the basics for identifying minerals. You don't often find pure minerals in the field, but when you do, there are certain easy tests you can apply to help narrow down your guesses. Over time, you'll get an "eye" for what you see and won't have to rely on these tips, but they're important when starting out. We'll begin by defining minerals and understanding their basic properties.

A mineral is a naturally occurring chemical compound. Kidney stones and clamshells don't count. To be a mineral, with few exceptions there must be a repeatable crystalline form at room temperature, and there must be a specific chemical formula, expressed as the ratio of elements normally present, as we showed above. Minerals can have two names: a common name and a chemical formula name. For example, calcite can be referred to as calcium carbonate, although it rarely is. Calcite is defined as a calcium atom, a carbon atom, and three oxygen atoms: $CaCO_3$. Its cousin dolomite has manganese substituting heavily for calcium, with the formula $CaMg(CO_3)_2$.

Impurities and substitutions make things interesting for so many minerals that we use terms such as "family" or "series." For example,

Disintegrating olivine-rich basalt colors this beach in Hawaii a pleasing green.

olivine is defined as a range: $(Mg,Fe)_2SiO_4$. It's a silicate, indicated by the single silicon atom attracting four oxygen atoms. The amount of magnesium and iron can vary, from forsterite, which has two magnesium atoms and no iron and is represented as Mg_2SiO_4, to fayalite, represented with two iron atoms: Fe_2SiO_4. In the field you'll just note the small olive green crystals, such as at the famed Green Sand Beach on the Big Island of Hawaii.

There are many more examples of this substitution pattern in minerals. Plagioclase feldspars form a continuous series from albite, rich in sodium, to anorthite, rich in calcium. Intermediate minerals in between include oligoclase, andesine, labradorite, and bytownite. Again, it's interesting to know of the series, but in the field you'll just note that feldspars are present.

Mineral Families

To bring order to the thousands of minerals identified and named to date, scientists have their own hacks. There are six main mineral families, described in table 2.

Table 2: Six Major Mineral Families, Each with Iron as an Example

Family	Description	Example
Oxides	Contain oxygen and one other element, but not silicon	Hematite $Fe2O_3$
Sulfides	Contain sulfur and one other element	Pyrite FeS_2
Chlorides (Halides)	Contain chlorine and one other element. Chlorine is most common, but fluorine, bromine, or iodine also work.	Molysite $FeCl_3$
Carbonates	Contain carbonate—CO_3—and one or more metals	Siderite $FeCO_3$
Sulfates	Contain sulfate—SO_4—and one or more metals and "n" water molecules	Melanterite $FeSO_4*nH_2O$
Silicates	Contain silicon and oxygen (quartz) or one or more metals	Fayalite Fe_2SiO_4

You can see that iron forms a mineral in each of the major classifications.

Mineral Properties

There are currently about 5,500 distinct minerals known, with around 5,312 approved by the International Mineralogical Association. ***Industrial minerals*** are important for commercial use, such as zeolites, used to help purify air, or fluorite, used as a flux to smelt metals. ***Ores*** are minerals with a high enough concentration of a single element (or elements) to be considered as a source for that element. For example, the zinc sulfide sphalerite is the most important ore mineral for zinc. ***Gems*** are minerals with value for jewelry because of their color, ease of use, rarity, or other characteristics. There are about thirty-five different gem minerals. Only about 150 minerals are considered important for collecting. There is, however, an entire field of "micromounts" where intrigued mineralogists search out thumbnail-sized specimens of unique and exotic specimens, which tend to be so small they are best appreciated under a microscope.

Eight elements account for over 98 percent of the Earth's crust by weight: oxygen, silicon, aluminum, iron, magnesium, calcium, sodium, and potassium. Oxygen and silicon are by far the two most important—oxygen composes 47 percent of the crust by weight; silicon accounts for

This tiny sliver of native silver in calcite from Kongsberg, Norway, looks great under a microscope.
RICE MUSEUM OF ROCKS AND MINERALS

28 percent. You'd expect that minerals with abundant oxygen, such as oxides, and minerals with abundant silica, or silicates, are all around us.

One key fact a practical geologist needs to understand is that an abundance of silica can be deadly. If you bang a hammer on too many silicates, you're liable to breath in an unhealthy amount of silica dust. That can cause tiny cuts and scars on your lungs, leading to silicosis. If you conduct any activity that involves hammering, cutting, sanding, or scribing on rocks for long periods of time, always wear a mask.

Let's talk about some of the important mineral characteristics next, to further refine the ability to identify and utilize them.

Luster

Luster refers to the way our eye interprets light interacting with the surface of a mineral (and some rocks). As you can imagine, luster is not easily defined, and there are no machines to measure it yet, although textile manufacturers are getting closer. It's a qualitative category, with a combination of reflectance and transparency. There are also gradations, or mixes. You'll want a specimen that is clean and untarnished, and you'll need good light.

You can divide luster into two major groups: metallic, and nonmetallic. After that, it gets crazy.

Hardness

Minerals have a distinctive and measurable resistance to scratching that you can test with some common household items, or you can purchase an expensive hardness test that can pinpoint a measurement. German mineralogist Friedrich Mohs developed the scale in 1812, but the concept that you can rank minerals by what they can scratch (and what can scratch them) has been around since Theophrastus in 300 BCE and Pliny the Elder in 77 CE.

Other hardness scales, such as the Vickers hardness test developed by a British engineering conglomerate, measure how far a diamond-tipped, pyramid-shaped penetrator can deform a surface. Variations include the Brinell and Rockwell procedures, but they all require expensive machinery in a laboratory setting. These scales are precise, but for rockhounds, simple tests are all you need.

Table 3: Mohs Hardness Scale

Number	Mineral (in bold)
1	**Talc**
2	**Gypsum**
2.5	*Fingernail*
3	**Calcite**
3.5	*Copper penny*
4	**Fluorite**
5	**Apatite**
5.5	*Glass plate, common knife*
6	**Orthoclase feldspar**
6.5	*Steel nail, steel file*
7	**Quartz**
8	**Topaz**
8.5	*Masonry drill bit*
9	**Corundum**
10	**Diamond**

You can put together your own hardness kit with very little expense. The entries written in italics in table 3 are commonly available and very helpful. The minerals, listed in bold, are the "official" test for that hardness number. (You can sometimes find mineral samples with impurities

You can create your own hardness kit from specimens and household items. From lower left: glass, fluorite, apatite, feldspar, quartz, a nail, and a steel blade.

that don't exactly work out to their official number.) Most rock shops have all the minerals you need to build your own kit. You don't really need a diamond, by the way—it's so hard, you can eliminate most minerals by scratching with a common ruby or topaz, both of which are less expensive than you might think.

Most of the time, you're going to be confused by minerals in the 4 to 7 range. If you have a nail, a steel file, a piece of glass, and a nice quartz crystal, you'll be able to work your way to a reasonable hardness measurement easily.

Streak

The streak test is very easy to use. The basic idea is to reduce a mineral to a powder and identify the color. To create a powder, you can use a piece of rough porcelain, which has a hardness of 7. Any hardware store that sells porcelain tile will have samples or broken pieces you can probably have for free, or you can purchase a few of the smaller "fiddlesticks" for your kit. More than one rockhound has turned over the cover of their toilet tank and streaked on the rough underside.

The truth is that while a streak plate is small, economical, and easy to use, there are so few minerals that produce reliable streaks that you will eventually learn to recognize them anyway.

Habit

A mineral's crystal habit refers to its tendency, or habit, of growing into a characteristic shape or form. Mineralogists refer to the crystal form as the outward appearance of its true geometry. However, in the field, you may find that the crystal habit is a general shape and takes into account irregularities. Usually, the atomic structure of the native crystal is the dominant force, but in some cases, the environment can play a factor.

Sometimes it seems as if there is a separate name for each mineral's habit, but after some practice, the different habits can help identify a mineral, or at least narrow its identification. However, quartz can occur in many different habits, so you'll need time to use appearance as your first clue. If you need more help, purchase a good mineral field guide, or look for more complete definitions with images online at sites such as dictionary.sensagent.com, geology.com, webmineral.com, and others.

Many white minerals look similar on first glance: Quartz, calcite, feldspar, and fluorite are all common at old mines. But quartz is hexagonal, calcite has characteristic rhombohedral shape; feldspar is also hexagonal but not has hard, and fluorite is typically cubic, like table salt.

Notice how the calcite crystals aren't perfectly square—they are rhombohedra, with distinctive angles.

Cleavage

Cleavage is the tendency for minerals to split reliably, along smooth planes. When minerals don't cleave, they fracture. Table 4 shows some of the more common cleavages.

Table 4: Mineral Cleavage Examples

Mineral	Crystal Habit	Cleavage
Halite	Cubic	Three directions at 90° angles
Calcite	Rhombohedral	Three directions, at 120° and 60° angles
Gypsum	Monoclinic	Cleavage in one direction
Muscovite	Monoclinic	Cleavage in one direction
Feldspar	Monoclinic or triclinic	Cleavage in two or three directions

This muscovite mica has extremely thin sheets that you can cleave off with a pocketknife.

Specific Gravity

When we say a mineral specimen is heavy, it has almost no meaning. To a strong man, the mineral may seem light as a feather. To an infant, the rock may be immovable. Rather than use relative terms such as "heavy," "massive," or "big," we want to be able to compare everything against a standard. The universal standard is pure water, which weighs 1 gram per cubic centimeter.

Therefore, we need two measurements: the absolute weight of the sample, in grams, and how much space it occupies, in cubic centimeters. Then we just do a little math.

One problem is that in nature, density often varies. Field geologists learn to expect how heavy a hand sample should be. You can hold the sample in your hand and let your hand drop slightly, then bring it up slowly. You'll feel gravity tug at the sample, and if it surprises you by being heavier or lighter than you expected, you've got a little more information. This is called the "heft" test. It's a very practical test to learn: Look at a specimen, quickly determine how heavy you expect it to be, and perform the heft test to verify your guess. In fact, that's the scientific method in a nutshell—observe, predict, experiment, and thus test your prediction.

Tenacity

The term "tenacity" applies to several characteristics. It refers to the way a mineral reacts to stress, such as crushing, bending, or breaking.

Table 5: Mineral Tenacity Examples

Term	Explanation	Examples
Brittle	When hammered or crushed, creates a powder or small chunks	Most minerals are brittle.
Sectile	Can be cut into sections with a knife, like wax	Gold, talc, gypsum
Ductile	Can be stretched	Most metals, but especially gold
Flexible, but elastic	Can be bent and springs back to original form	Mica
Flexible, but inelastic	Can be bent but stays in the bent position	Copper
Malleable	Can be flattened by pounding with a hammer	Most metals, but gold, silver, and copper are notable.

One of the most tenacious minerals you'll encounter in the field is nephrite jade. Mayan and Aztec warriors favored weapons using jade at the business end of a war club because it didn't shatter when used. Sometimes you can't even chip it with a well-struck hammer blow in the field. Compare that to obsidian, which when flaked to a very sharp edge, tends to break easily, if not outright shatter.

Color

You can use color to identify a few common minerals. Unfortunately, many minerals come in a variety of colors, so color isn't a perfect identification tool. Quartz can be clear or milky white, for example. When colored purple, quartz is called amethyst. When black, it's called smoky quartz. When colored with chlorite, quartz can turn green; colored with rutile, it can be golden yellow. Fortunately, it usually still has a hexagonal habit.

Exposure to air, heat, and sunlight can also darken, lighten, or fade a mineral. Iron can rust, and it can color a quartz crystal dull red or brown. Some metals, such as silver or copper, tarnish. If you leave an amethyst out in the sun, the purple can fade. The same goes for precious opal—exposing a good sample to sunlight can ruin it.

The challenge comes when minerals mix freely. Blue azurite and green malachite often mix in swirls, as do chrysoprase and variscite. Turquoise can occur as thin coatings.

Interesting variscite specimen from Utah's famed Clay Canyon
RICE MUSEUM OF ROCKS AND MINERALS

Luminescence

There are several different types of luminescence in minerals, described in table 6.

Table 6: Mineral Luminescence Examples

Category	Explanation	Examples
Fluorescent	When subjected to ultraviolet rays, either shortwave or longwave, some minerals absorb the energy and re-emit it at a lower wavelength.	Calcite—red, pink, light green Halite—red Dolomite—red Opal—green Fluorite—purple Gypsum—blue
Phosphorescent	Absorbs ultraviolet light and re-emits visible light afterward.	Calcite
Thermoluminescent	Applying heat spurs the mineral to glow.	Calcite, feldspar, fluorite
Triboluminescent	Under mechanical stress, including scratching, rubbing, or striking, changes to the crystal lattice cause light to emit.	Sphalerite Wint-O-Green Lifesavers

Variety of fluorescent minerals, many from the famed Sterling Hill mine in Franklin, New Jersey
RICE MUSEUM OF ROCKS AND MINERALS

UV lights are a great addition to your field kit. Not only can you detect calcite, you can also find scorpions, which pack a lot of calcite in their exoskeletons.

Radioactivity

Back in the 1950s, there was a serious rush for uranium ores in many western US mineral districts. Young men armed with Geiger counters swept through the deserts hoping for the rapid clicking of a strong radioactive signal. Most probably died of cancer if they stuck at it for very long. Very few collectors have anything to do with such minerals today; such specimens tend to emit radon in their cabinets, which can be deadly.

Geiger counters are not terribly expensive, and there are reports of less-sensitive devices that plug into a smartphone. Photosensitive paper will also turn cloudy around such minerals.

Many radioactive minerals are distinctively neon yellow or green, such as carnotite (CAR-no-tite), autunite (ah-TOON-ite), or uranophane (you-RAIN-o-fane). Others are not, and there are about eighty-one minerals with uranium as a component; most are extremely rare. Carnotite is interesting because it occurs in sandstones and sometimes replaces petrified wood. The formula is $K_2(UO_2)_2(VO_4)_2{*}3H_2O$, so it is also a source of vanadium.

This solid old Geiger counter emits a telltale rapid clicking when it detects strong radiation.

Magnetism

Some minerals react weakly or strongly when placed near a magnet. It's not a foolproof test, as impurities and contamination with other minerals can interfere. Magnetism is an excellent test for meteorites, as almost all varieties are magnetic to some degree.

Acid Reaction

Calcite (and its relative, aragonite), or calcium carbonate, is the main ingredient in the shells that make up limestone and chalk. You can test for it by using an acid, such as white vinegar, which is a diluted acetic acid.

Here's the equation for what happens when common vinegar reacts with calcite:

$$CaCO_3 + 2CH_3CO_2H = Ca(CH_3CO_2)_2 + H_2O + CO_2$$

If you take the time to count up the atoms, the equation balances. One calcium carbonate molecule plus two molecules of acetic acid equals calcium acetate, which forms a residue at the bottom of your container, plus water and carbon dioxide. The carbon dioxide is what forms the bubbles.

One problem in the field is that dolomite and limestone are often found together. Dolomite has magnesium in it—the chemical formula is $CaMg(CO_3)_2$. Geologists sometimes call dolomite a double carbonate. Unless you grind dolomite into a fine powder, it won't fizz at room temperature as limestone does, and sometimes you need a stronger acid and some heat to test for it.

Mineral Keys

Now that you know how to test for various properties, you can more easily use a key to figure out what you have. The key works as a series of questions, and you usually work your way through it by figuring out if you have a metal or not and then use color, habit, streak, hardness, and density to narrow down the search. There are apps for keying out rocks and minerals, and many advanced mineralogy books include keys.

THE THREE ROCK TYPES

Now that we understand minerals, we can advance to rocks, which are composed of minerals in countless ways. In this section we'll discuss the three main rock types but try to keep the information relevant to newcomers as much as possible.

Fortunately, just about everyone knows the three main rock types: igneous, metamorphic, and sedimentary. For a refresher, you can think about how rocks form and how the Earth recycles them.

Let's start in the kitchen: Think of the pre-made chocolate chip cookie dough you dig out of a container and place on a baking sheet. The "liquid" cookie dough is similar to a hot igneous rock that comes out of a volcano. Once baked, the dough changes somewhat; it hardens—like a metamorphic rock. Then, if you eat the cookies until all that remains are the crumbs, those are the same as sedimentary rocks. Now you have a tasty metaphor for the rock cycle.

The three rock types, represented by chocolate chip cookie dough, baked cookies, and the remnant crumbs.

The ingredients are basically the same during each stage, but the chemistry changes with heat and water loss, and mechanical forces accelerate erosion. Theoretically, instead of eating the cookies, you could crush them all up, add the water that was in the original dough, and reheat into a new cookie to make a complete cycle. Or just eat the cookies and use your imagination.

Most rocks at the surface of the earth are sedimentary rocks called alluvium (from the Latin *alluvius*; from *alluere*, "to wash against"). Alluvium is often referred to as "dirt," but to geologists, the word "dirt" has no real meaning. Alluvium is loose soil or sediment that has been eroded, reshaped by water in some form, and redeposited in a nonmarine setting. It is typically not considered rock yet, because it hasn't been cemented together. If you look at a geology map, the most common color is often yellow, signifying the Quaternary period, and the term "Qal" for "Quaternary alluvium" is a common label.

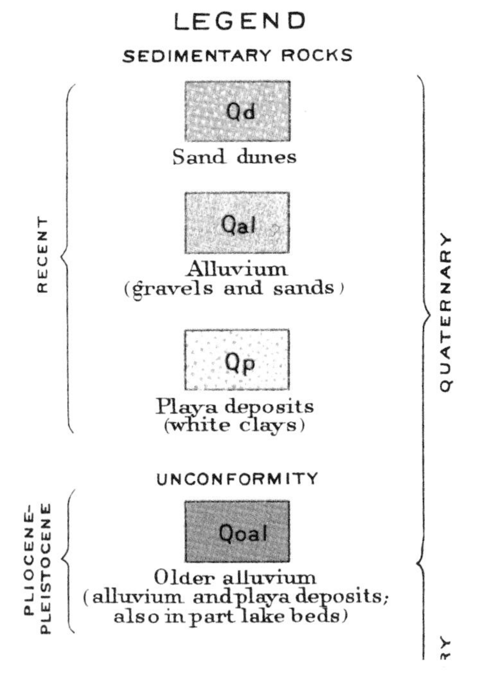

The top section of the map key for a geology map, showing the placement of Quaternary alluvium

Igneous, sedimentary, and metamorphic rocks are all present in the Earth's crust, but they are not evenly distributed. Most of the Earth's crust—95 percent—consists of igneous and metamorphic rock. Sedimentary rock, which forms a thin covering on the Earth's surface, makes up only 5 percent of the crust, yet 75 percent of the rocks at the surface of the crust are sedimentary—especially if you include the soil cover.

Sedimentary rocks such as sandstone, mudstone, and limestone are made up of sediments from erosion or, in the case of limestone, can form chemically as tiny organisms die and settle to the bottom of a lake, lagoon, or ocean basin.

Igneous rocks are more complex. Imagine a giant granite blob slowly cooling underground, such as the great Sierra Nevada Batholith before it became the Sierra Nevada mountain range. As it cools, most of the liquid materials form into common minerals. Aluminum, silicon, calcium, magnesium, oxygen, iron, and hydrogen will join as feldspars, pyroxenes, or even simple quartz crystals, and make up most of the granite. However, there is often a lot of liquid left over that is still hot and still looking for another element to hook up with. This isn't plain water—it's more likely to be a hot quartz-rich brew, carrying all the leftovers.

Picture a typical high school dance. At first the boys are on one side of the room, the girls on the other. All the popular kids pair up first, similar to the most common elements in the crust. At the end of several songs, all that's left on the sidelines are the "exotic" kids that are just harder to find matches for. It's the same way with chemicals—silica, oxygen, iron, calcium, aluminum, and sodium are popular and find matches quickly. What's left tends to be much more interesting—gold, silver, copper, platinum, and other precious metals, for example. There is often quite a lot of excess silica or quartz. The residuals remain in the quartz, looking for a match with sulfur, oxygen, and so on. There are 118 elements in the periodic table, but only 8 are significant. Think about that—110 elements are grouped into the "All Others" row in table 7.

Table 7: Chemical Composition of the Earth

Element	Percentage
Iron	32.1
Oxygen	30.1
Silicon	15.1
Magnesium	13.9
Sulfur, Nickel, Calcium, and Aluminum	7.6
All Others	1.2

Igneous Rocks Explained

There are two main kinds of igneous rocks—intrusive and extrusive. Igneous rocks either break through and erupt dramatically, as extrusive rocks, or they stay underground, as intrusive rocks. Think of extrusive rocks as bold and boisterous, while intrusive rocks are shy and quiet. That part is simple to understand, but there's a little more to consider.

Intrusives

For igneous rocks to form, they need a source, usually referred to as a magma chamber. Think back to your first exposure to the concept of plate tectonics: When dense, thin oceanic plates sink under fatter, lighter continental plates, the material is buried deep in the Earth's crust. If there is enough heat and pressure, the plate material melts completely and tends to want to rise, similar to the blobs in a classic lava lamp.

Intrusive rocks don't erupt with a big explosion or a giant, glowing red lava flow from a volcano. Instead, they rise from the depths of the Earth's crust but don't make it all the way through. They just sit there and cool in place, usually very slowly, so they tend to have big individual crystals. The faster molten rock cools, the smaller the individual crystals, as they don't have time to align with similar molecules and form big clusters.

Common intrusive rocks include granite, the most common, and diorite, also common, with the intermediate granodiorite in between. There is also a rock known as a monzonite, and several others. To professional geologists, granite and diorite are two vastly different kinds of intrusive rock, but they're related enough that a practical geologist can get away with shorthand and just lump them together as granite and move on.

What's interesting is that if a diorite rock erupted as a lava flow, it would then be an andesite. If a granite intrusion erupted, it would likely be classified as rhyolite. Granodiorite that erupted would be called a dacite. The chemistry is the same, but the way it forms makes the difference. Most intrusives have larger crystals, while extrusives are finer-grained.

The most interesting intrusive rock for rockhounds is pegmatite. The word derives from the Greek *pegnymi*, meaning "to bind together"; think of a quilt with large pieces. Pegmatites have the same composition as granite—with plenty of quartz, feldspar, and mica—but they cooled very slowly and often formed pockets due to faulting, allowing large crystals to form. Pegmatites sometimes host interesting minerals such as garnet, fluorite, and corundum, as well as gemstones such as aquamarine, tourmaline, and topaz. Some pegmatites house rare earth elements such as tantalum and niobium, vital for cell phones and tablets.

Yosemite's famed Half Dome is an impressive chunk of granodiorite.

There are other intrusive igneous rocks, such as monzonite and gabbro. Many rockhounds use "granite" to refer to just about all intrusive rocks with a salt-and-pepper appearance.

Extrusives

Rocks that erupt, such as basalt, rhyolite, and andesite, usually cool quickly and so are normally fine-grained. Basalt is by far the most common extrusive igneous rock. Many of the world's major volcanic events were flood basalts that erupted for several miles along rifts in the Earth's crust. Later on we'll talk about some major extinction events in Earth's history that may have been associated with giant eruptions.

For the most part, extrusive rocks come from three methods:

1. **Divergent plate boundaries** that move away from each other, such as a rift or where the oceanic plates are opening up.
2. **Convergent boundaries,** also called subduction zones, where the plates come together and one plate scrapes under the other one.
3. **Hot spots,** where a plume of hot magma stays in one place but the plate moves across it. The Hawaiian Islands sit atop a hot spot, as does the Yellowstone Caldera.

When rhyolite lava flows cool extremely fast, they don't form minerals—they turn into a glass, called obsidian. There is no mineral structure to obsidian, which explains why it fractures so neatly for making arrowheads. Obsidian weathers faster than normal rhyolite, and there is little evidence of obsidian older than a few million years.

Understanding Volcanoes

Returning to our cooking metaphors, there are two recipes you can try that simulate igneous rocks. The first is a soufflé. If you want to practice some igneous geology, find a good recipe and bake a soufflé, but without trying real hard to prevent it from collapsing. Many recipes concentrate on monitoring the temperature precisely, using care when folding in the egg whites, not beating in too much air, or adding an acid such as lemon juice. Ignore all that, and use a round, flat bowl. Cook as normal, and watch the mix rise in the oven. As soon as you take it out, imagine the Yellowstone Caldera bulging in a similar way. As the steam escapes and the soufflé sinks, you have a perfect metaphor for a collapsing caldera.

Calderas are super volcanoes, sometimes 30 miles across or more. They can be extremely dangerous, blasting out enormous quantities of lava and ash. Nonexplosive calderas, such the one around Hawaii's Kilauea volcano, pour out lava with much less drama. The reason there is less drama is because of the amount of quartz present. If the magma is rich in quartz, the flows don't pour out easily. Common basalt has less quartz and flows better. The two main names for lava both come from Hawaii:

1. A'a (pronounced ah-ah) is coarse, blocky lava.
2. Pahoehoe (pronounced puh-HOY-hoy) is thin and ropy.

Note that these are textural terms. An individual lava flow can exhibit both textures—runny and ropy in the middle; blocky at the margins.

Basalt from the south side of Kilauea, showing ropy pahoehoe lava in the middle and blocky a'a lava around it.

As of 2021, Iceland's Fagradalsfjall volcano is thrilling viewers with long, streaming orange lava flows.

If you want to learn more about eruptions, there's an app for that. A good one is "Volcanoes and Earthquakes" from VolcanoDiscovery, which is highly rated and full of interactive features for tracking active volcanoes and the latest earthquakes worldwide.

Here's a second kitchen metaphor for understanding lava tubes, which are long, narrow caves in volcanic terrain. Consider a tasty lava cake. To have a good lava cake, you usually have a hot, cooked pastry

Iceland continually creates fresh volcanic rock, such as this still-smoking 2021 flow. SHARON MCLEOD

outside with a rich, molten interior. When you cut into it, the molten interior flows out—and you can now imagine how lava tubes form. If a runny lava flow fills in a creek or river, the lava often cools first at the top, forming a crust, and at the bottom, where it contacts other rocks. Inside, the interior of the lava flow may still be hot and runny, similar to a lava cake. If the molten lava starts flowing downhill, following the creek bed while the top and bottom are cooled and solid, you get a perfect situation to form a lava tube.

If you do start to take notice of all the volcanic activity around you, you'll come to realize that our Earth is young and restless. There are some volcanoes that erupt continuously, such as Stromboli, the "Lighthouse of the Mediterranean." Other volcanoes save their power up and spring to life with massive eruptions. If you were building a new homestead, you'd want to steer well clear of the volcanoes listed in table 8.

Before moving on, we should mention one more kind of volcanic rock—volcanic ash. This material is the debris that volcanoes and calderas blast into the air. Geologists use the term "tephra" (pronounced TEFF-ra) for fragmented material ejected during an eruption. (Another term is "pyroclasts," or "fire pieces.") Tephra is composed of three main types of material:

Table 8. Significant and Recent Volcanic Events

#	Name	Location	Recent	Notes
1.	Kilauea	Hawaii	2018	Very active since 1983
2.	Mount Etna	Italy	2017	Continuous activity for 3,500 years
3.	Piton de la Fournaise	Reunion Island	2018	Continuous activity since the 1700s
4.	Mount Nyiragongo	Democratic Republic of the Congo	2002	Active lava lake at summit
5.	Stromboli	Italy	2018	Continuous "Lighthouse of the Mediterranean" for 2,000 years
6.	Santa María	Guatemala	2013	Massive eruption in 1902
7.	Mount Yasur	Vanuatu	2013	"Lighthouse of the Pacific"—activity for last 800 years
8.	Láscar	Chile	2007	Major event in 1993
9.	Sangay	Ecuador	2013	Major event in 1628
10.	Mount St. Helens	Washington, USA	1980	Dome-building continues

1. Ash, made up of smaller particles less than 2 millimeters in diameter
2. Lapilli, or volcanic cinders, between 2 and 64 millimeters
3. Bombs (or blocks), which are larger than 64 millimeters

Volcanic ash piles up around and downwind from explosive volcanic eruptions when the air dissolved in molten lava expands rapidly as it reaches the surface. The air escapes with such force that the surging lava basically explodes, boiling into the atmosphere where it cools and drifts in the wind.

One frightening form of volcanic ash is the "*nuee ardente*" (NOY arden-TEY), which is French for "glowing cloud." The term is used for a very, very hot cloud of volcanic gas and ash. The gas cloud can't rise very high, so it burns down the flanks of a volcano at greater than 50 miles per hour and can be as hot as 1,000°C. It's not hard to imagine how dangerous that is. When Mount Vesuvius buried Pompeii in 79 CE, it belched out a massive cloud of superheated tephra, pumice, molten rock, and noxious gasses. The resulting pyroclastic debris and ash fall buried the area in several feet of hot ash, preserving many doomed inhabitants in cast molds.

Ash deposits from the ancestral Yellowstone Caldera reached 1 foot in depth nearly 1,000 miles away, preserving as fossils a herd of rhinoceros at Nebraska's Ashfall Fossil Beds State Historical Park. The John Day Fossil Beds in central Oregon are also the result of volcanic eruptions. Ash falls tend to be so rapid and catastrophic that they are a key resource for fossil hunters.

Similarly, ash beds (also known as "tuff beds") tend to do a great job of preserving petrified wood and fossil leaves. Not only that, but the silica-rich solutions washing through recent ash falls often form layers of opal and agate—especially when repeated sequences of lava flows and ash falls create a layer cake of volcanic rock. The lava flows seal the ash layers, so circulating fluids encounter traps. Thunder eggs and geodes occupy mixed beds of ash and lava as well.

Metamorphic Rocks Explained

Metamorphic rocks form when igneous rocks or sedimentary rocks undergo heat and pressure, changing their form. It's not hard to understand what's going on—heat and pressure are powerful forces. Think back to the kitchen metaphors we've been using and add in the useful pressure cooker once employed to can fruits and vegetables. Pressure tends to accelerate the effects of heat, reducing the time needed for transformation.

The best guess is that about 12 percent of the Earth's land surface is metamorphic rock, and, presumably, most of the Earth's crust is metamorphic. There are two main methods for creating a metamorphic rock. The most significant are regional processes, such as plate tectonics, which cause plates to collide with each other or slide under, as discussed earlier. The other process, much more localized, happens when a large igneous intrusion pushes its way up, heating up the rocks along its margins as it rises. This is called contact metamorphism. Fortunately, the effects are the same, and there is no difference in naming a limestone that transformed into marble, whether it was caused by plate tectonics or a rising intrusion.

When rocks are metamorphosed, minerals can reform in new combinations. Here's another metaphor from the kitchen to help you think about the way metamorphism creates interesting crystals. If you've ever had a very fine, aged cheddar cheese, you've probably noticed there are small crystals that add grittiness. These are either simple calcium lactate crystals in younger cheeses or amino acid crystals in finer aged cheeses. The point is that as the cheese aged, it accumulated crystals in pockets. They weren't there to start with but developed with aging. In the same way, we can get garnets, staurolite, and kyanite in metamorphic rocks.

Table 9: Metamorphism and Temperature Ranges (pressure omitted)

	Very Low-Grade	Low-Grade	Medium-Grade	High-Grade Metamorphism	
Temp range	150° to 300°C	350 to 450°C	450 to 550°C	Above 550°C	
Rock type	Slate	Phyllite	Schist	Gneiss	Migmatite
Minerals	Chlorite	Muscovite	Garnet → Staurolite → Kyanite →	Sillimanite	
Foliation	Tight foliation				Coarse foliation
Crystal size	Small crystals				Large crystals

Table 9 is a generalization, and pressure is left out to simplify things. In general, foliation (from the Greek *folios*, for "leaves") increases to the right. So does crystal size. Garnet doesn't stop forming after medium-grade metamorphism; if the minerals are right, you can get garnets in a gneiss. There isn't a well-understood border between many of these grades, either. The boundary between schists and phyllites (FILL-ites) is arbitrary, for example. However, the boundary between kyanite (KIGH-a-nite) and sillimanite is firm and marks the transition from schist to gneiss.

Some metamorphic minerals are of interest to rockhounds. Staurolite can form attractive twins in an "X" or a cross, such as in Fannin County, Georgia, or at Fairy Stone State Park in Virginia. Garnets are always fun to pan out, and when they get bigger or show few fractures, they achieve gem status. The USDA Forest Service operates a fee-dig area at Emerald Creek in northern Idaho, and rockhounds come from all over the world to pan special "star garnets," which have inclusions that form a four- or six-pointed star when polished correctly.

Kyanite is an attractive blue mineral that forms long blades. It's also one of the few minerals that have a different hardness depending on which direction you try to scratch it. Sillimanite takes a nice polish and appears to have tiny fibers; another name for it is fibrolite. Interestingly, sillimanite, kyanite, and andalusite share the same chemical formula—they're aluminum silicates, or aluminosilicates. However, they have different shapes, so they're called polymorphs.

Foliation

Many metamorphic rocks have lines in them, such as when sandstone is formed into a schist. Those lines, called foliation, may represent the original straight lines of a sedimentary rock when it was deposited, but they are now tighter and wavier. Depending on how much cooking is going on, metamorphic rocks grade from phyllite and schist, which are only lightly folded and toasted, to wavy, banded gneiss, which has undergone severe heat and pressure.

Table 10: How Metamorphic Rocks Change with Temperature

Starting rock	150–300°C	300–450°C	450–550°C	Above 550°C
Mudstone	Slate	Phyllite	Schist	Gneiss
Sandstone	No change	Little change	Quartzite	
Limestone	Little change	Marble		
Basalt	Chlorite schist (greenstone)		Amphibole gneiss	
Granite	No change			Granite gneiss

ADAPTED FROM KARLA PANCHUK, *PHYSICAL GEOLOGY*, 2ND EDITION. CREATIVE COMMONS.

What you should notice is the way the slices apparently moved around. In metamorphic rocks, these lines are called foliation. They're an important way to tell the difference between sedimentary rocks, petrified wood, and metamorphic rocks. Petrified wood is often square and stubby, and the lines are usually straight and uniform. Schists tend to have wavy lines that thin and thicken. Sedimentary rocks usually have very fine lines.

Finally, marble is a primary metamorphic rock used by architects for stately monuments and buildings. One of the premier marble deposits in North America is the Yule marble from Colorado. It's a gleaming, massive white rock, is very uniform in appearance, is almost pure calcite,

The marble from this old mine in Nevada had too many cracks and fissures for commercial use.

and has a beautiful, luminous surface when polished. The Washington Monument, the Lincoln Memorial, and the Tomb of the Unknown Soldier in Washington, DC, were all created with Yule marble.

Contrast the Yule marble with an American version of the famed Carrara marble that Michelangelo fashioned into famous statues such as *David*. In 1904 prospectors located what they believed was a large and valuable marble deposit in the hills southeast of Beatty, Nevada. Amid much fanfare and not a little speculation, city founders laid out a townsite, opened a post office, induced a railroad to lay tracks to the mine, and even commissioned a water fountain in the desert town. The bloom soon faded. While the famed Italian marble is uniform and frequently yields large blocks unblemished by cracks and fissures, the Nevada version soon proved unsuited for commercial use. The remains of the quarry are a testament to the vagaries of natural processes.

Serpentine is a common metamorphic rock that carves easily. It tends to have asbestos in it, so you must take care to avoid the dust, but it takes a polish and is easy to work.

Sedimentary Rocks Explained

The most common rocks at the surface of the Earth's crust are sedimentary rocks, which include alluvium (soil, or dirt) as well as sandstone, siltstone, limestone, and other material. There are four main methods for creating sedimentary rocks:

1. **Clasts:** Accumulating crumbled (clastic) rock material, from tiny grains to large boulders; examples are mudstone, sandstone, and conglomerates.
2. **Chemicals:** Accumulating evaporites; examples are gypsum and halite (rock salt).
3. **Biochemicals:** Accumulating material with help from organisms; examples are coal, limestone, and chert.
4. **Other:** A catchall for volcanic ash deposits that rumble down the slopes of volcanoes or falls from the air; impacts from large meteors.

Geologists can tell a lot about the clasts that make up a sedimentary rock just by seeing how rounded they are. You can assume that rocks such as breccias, which have lots of jagged edges and sharp angles, haven't moved far. Rocks that are rounded and smooth probably moved in water some distance from where they started. Larger boulders probably haven't moved far; most major river systems dump mostly sand at their mouths.

Key Observations About Sedimentary Rocks

Many of the key ideas we use to understand sedimentary rocks come from Nicolas Steno (1638–1686), a keen observer who described four basic laws:

1. **Law of Superposition:** In a normal sequence, the rocks at the top of an outcrop are younger than the rocks at the bottom. There are numerous examples where rocks have been flipped so that the older rocks are on top, such as around impact craters or in areas of intense folding. However, in general, as a valley floor fills in or as an ocean bay builds up sediments, the oldest rocks are always at the bottom.
2. **Law of Original Horizontality:** In most areas where rocks are deposited slowly but surely over time, the layers start out flat. This is especially true in lakes or oceanic basins. Sediments fall to the bottom of a lake and build up in layers that are always flat on top. There are exceptions—in river deltas, for example, the layers may build at an angle as a river covers a valley floor. Sand dunes are another example where layers don't start out horizontal. However, in general, sedimentary rocks start out flat and only tilt after folds, faults, and other tectonic forces shove them around.

3. **Law of Crosscutting Relationships:** We often observe rocks that host veins of material. We have to assume that the host rocks are older than the material that was injected into them. Crosscutting veins had to have entered the picture after the original rocks formed.
4. **Law of Lateral Continuity:** We know that rocks are laid down over large areas; therefore, we can assume that they would continue unless they encounter some form of a natural barrier, or if they are broken up by later events. We sometimes observe similar rocks on both sides of a valley, for example, with a wide gap in between where erosion has removed the rock that would otherwise appear.

One of the common sedimentary rock groups is the evaporites. As the name implies, these are the crystals left behind when water evaporates. Gypsum is a common evaporite; its chemical composition is $CaSO_4 * 2H_2O$. The chemical name is calcium sulfate dihydrate; there are water molecules locked up in the gypsum, accounting for its softness. If gypsum is buried for enough time, and thus subjected to heat and pressure, gypsum alabaster forms.

Gypsum crystals from a deposit in Nevada

A common evaporite is rock salt, or halite. There are a couple of interesting concepts here. First, there is the idea of "saturation." You can dissolve a lot of salt in water if the temperature rises.

The Dead Sea in Israel is said to be "supersaturated" with salt due to water temperatures that reach 90°F, and in some places perfect tiny cubes of halite form in the gravels. In Utah's Great Salt Lake, you often see salt-encrusted sticks and rocks along the shore. Each time a wave encounters a stick or rock, a little more salt is added to the already present crystals, and the area can continue to grow.

Other related evaporite minerals, called halides, come from compounds with ions such as fluoride (F–), chloride (Cl–), bromide (Br–), iodide (I–), and others. Sylvite is the mineral potassium chloride, or KCl. Other evaporites include carbonates, such as calcite ($CaCO_3$).

Conglomerates are masses of pebbles that are cemented together. Some are very hard; others aren't. Hardness usually depends on how much calcium was present in the liquid that bound up the pebbles—the more calcium, or lime, the harder the rock.

We've now scratched the surface of the three rock types and learned how they are of interest to rockhounds. The truth is that you'll see a lot more rocks in your lifetime than you will ever see minerals, so keep that in mind and try to learn the most common rocks.

THE EARTH AND ITS CORE

Imagine the Earth as a big piece of round layered candy, slowly spinning from west to east if viewed from the side or counterclockwise if viewed from above. The region at the center, the inner core, is thought to be solid nickel-iron, followed by a molten outer core, and then a gooey mantle, with an upper and lower zone. On top of all that is the crust, the rocky skin, which is only about 100 kilometers thick at the most. The crust under the ocean can be as thin as 5 kilometers. We rarely see rocks from the mantle; they're dark, heavy rocks that require some unique events before we can collect them.

A tremendous amount of heat is generated from the core; it has to go somewhere, so it rises. You've seen water at a rolling boil in a pot on the stove, so you know how heat moves through a liquid. But imagine how heat would move through melted rock that has the property of liquid plastic or toothpaste. These plumes of rolling heat move slowly outward through the mantle and push continents around at the crust, spawning volcanoes and creating earthquakes.

The deepest gold mine in the world is the Mponeng Mine near Johannesburg, South Africa. At its greatest depth, 4 kilometers (about

2.5 miles), temperatures reach 150°F according to generalkinetaics .com. Operators must pump down an ice slurry so the miners can work safely.

Know Your Crust

The thick continental crust is less dense than oceanic crust. Continental crust is about 35 to 40 kilometers (22 to 25 miles) thick, versus the average oceanic crust thickness of about 7 to 10 kilometers (4 to 6 miles). A continental crust currently covers about 40 percent of the Earth's surface, but that percentage has changed a lot over the ages.

To picture how the Earth recycles rocks, consider the Ring of Fire, consisting of hundreds of volcanoes surrounding the Pacific Ocean. In the middle are rift zones, where new material erupts beneath the ocean. At the edge of tectonic plates, the rocks slide underneath and will eventually end up so far down that there is too much heat and pressure for them to remain solid.

Some material may not get that deep and may simply change into a metamorphic rock, as the pressure and heat cook the rock but doesn't melt it. Other times, the rocks go deeper and melt into magma; the magma wants to rise like a bubble, burning its way up through the solid crust. As the plates continue to collide and more material is fed into the system, the "blob" can eventually swell up enough to break through the crust as a volcano. Alternatively, it may not break through, and instead "freezes" in place.

During many periods of the Earth's formation, there was only one large continent. About 335 million years ago, geologists believe that the entire continental crust was swept up into one big island as chunks combined with other masses to form the supercontinent Pangaea. Based on fossil evidence and other studies, it is believed that Pangaea probably started breaking up about 176 million years ago.

Geologic Time

The Earth is both young and old at the same time. It is young because the internal "furnace" is still burning and we have volcanos erupting all the time. On Mars, geologists believe that tectonic forces have almost completely subsided; there is just one super volcano, Olympus Mons, which towers about 72,000 feet tall—2.5 times the height of Mount Everest. It reached that great height because it sits over a hot spot that never moves, unlike Yellowstone or the Hawaiian Islands.

However, the Earth is also incredibly old. Using radiometric dating, which relies on the decay of certain radioactive elements, scientists have determined that the Earth is about 4.54 billion years old.

Table 11 shows only the ages of the major time periods.

Table 11: Broad Overview of Geologic Time Periods

Geologic Label	Range of Years	Important Events
Cenozoic era	66 million years ago–present	Rise of mammals
Mesozoic era	252 million–66 million years ago	Age of dinosaurs
Paleozoic era	541 million–252 million years ago	Explosion of complex life forms
Precambrian super eon	4.54 billion–541 million years ago	Formation of the Earth

Naturally, this can be broken down a lot more. Just consider the Cenozoic, the era we are most familiar with because we live in it. As shown in table 12, the Cenozoic is divided into three periods: the Paleogene, Neogene, and Quaternary; and seven epochs: the Paleocene, Eocene, Oligocene, Miocene, Pliocene, Pleistocene, and Holocene. The Quaternary period was officially recognized by the International Commission on Stratigraphy in June 2009. In 2004 the Tertiary period was officially replaced by the Paleogene and Neogene periods.

Table 12: Breakdown of the Cenozoic Era

Era	Periods	Epochs
Cenozoic	Quaternary	Holocene (present–11,800 years ago) Pleistocene (11,800–2.6 million years ago)
	Neogene	Pliocene (2.6 million–5.3 million) Miocene (5.3 million–23 million)
	Paleogene	Oligocene (23 million–34 million) Eocene (34 million–56 million) Paleocene (56 million–66 million)

Table 13 describes the Holocene in greater detail.

Table 13: Stages of the Holocene Epoch

Epoch	Stage	Time Period	Notes
Holocene	Meghalayan	4,200 years ago to present	Named for Krem Mawmluh Cave in India's Meghalaya State
	Northgrippian	8,200–4,200 years ago	Named for the NorthGRIP Ice Core Project
	Greenlandian	11,650–8,200 years ago	Glacial retreat

Geologists studying time use the term "BP," or "Before Present," rather than "years ago" or "Before Christ" (BC). But since that marker moves, the year 1950 is used for "present." Another term is Before the Common Era—BCE, which replaces references to Christianity and is thus more acceptable globally. That means you can usually convert between BCE, BC, or BP by adding or subtracting roughly 2,000 years.

Notice how a geology table starts with the oldest information at the bottom, following Steno's Law of Superposition. Again, you can think of the Earth as a layer cake, with someone adding new layers of frosting, sprinkles, and other thin layers all the time. Geologists work out the layers and name them by drawing a "stratigraphic column" or map of the rock strata they see or believe to be there. The layer on top will almost always be the youngest.

What happens all the time, however, is there are gaps in the stratigraphic column. Geologists call these "unconformities" because in a perfect world, there would be no gaps. Millions of years can go by between deposition of more sedimentary rock, or between volcanic eruptions. In other cases, the cake may be tilted by earthquakes and other

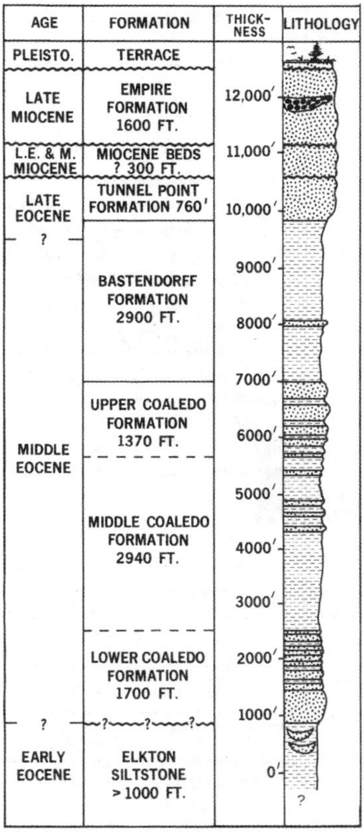

A stratigraphic column depicts the different rock formations, shows their age based on their relationship to one another, and includes a reference to the ability of the rock to withstand weathering.

forces, causing some layers to be laid down at different angles than both the earlier rocks and the later ones. In addition, the sheer weight of younger rocks piled up higher and higher can cause changes in the rocks below.

A New Time Period?

Soviet scientists used the term "Anthropocene" as early as the 1960s to refer to the current geological period, and ecologist Eugene F. Stoermer picked up on it by the 1980s. By 2000 atmospheric chemist Paul J. Crutzen described the impact of humans particularly in the atmosphere as so significant that a new epoch was needed to describe it.

"Anthropocene" is a combination of *anthropo-* from the ancient Greek *Anthropos*, meaning "human" and *-cene* from *kainos*, meaning "new" or "recent." The Anthropocene is proposed to mark the current time period, but where it should begin is still subject to vigorous debate. Various start dates for the Anthropocene range from the beginning of the Agricultural Revolution (12,000–15,000 years ago) to as recently as the 1960s. The peak in radioactive fallout from testing nuclear weapons during the 1950s has been more favored than others, locating a possible beginning of the Anthropocene to the detonation of the first atomic bomb in 1945, or the Partial Nuclear Test Ban Treaty of 1963.

Those arguing for earlier dates posit that, based on geologic evidence, the proposed Anthropocene may have begun as early as 14,000–15,000 years BP; this has led other scientists to suggest that "the onset of the Anthropocene should be extended back many thousand years"; this would make the Anthropocene essentially synonymous with the current term "Holocene." So far the Anthropocene has not gained formal status, however.

Another proposed addition to the geologic calendar is the Homogenocene, from the old Greek *homo* ("same"), *geno* ("kind"), plus *kainos* ("new"). The idea is that biodiversity is plummeting and global systems are becoming similar as invasive species mix and mingle and agriculture exploits similar crops and livestock. In 1999 entomologist Michael Samways wrote an article for the *Journal of Insect Conservation* titled "Translocating fauna to foreign lands: Here comes the Homogenocene." Subsequent authors have picked up the term, but as yet nothing official has been decreed.

CHAPTER 3

IMMEDIATE SURVIVAL

The "Rule of Threes" is a survival tip to guide your decision making in the wild under extreme conditions. In *The Survival Handbook* (DK Publishing, 2012), Colin Towell, a British Special Air Service (SAS) combat survival instructor, expands on the Rule of Threes and builds out the decision-tree as follows:

3 seconds	***to get your head straight and make good decisions***
3 minutes	***to get air before dying from lack of oxygen***
3 hours	***to get shelter before dying of exposure***
3 days	***to get water before dying of thirst***
3 weeks	***to get food before dying of hunger***

So in terms of immediate survival, you need to be alert, breathing good air, and out of the elements. Let's first look at keeping a clear head.

REACTION TIME HACK

In certain situations, your senses can be dull and sluggish. You need some kind of "pick-me-up" that is readily available. Certain botanicals can solve the problem; garlic, mint, lemon balm, rosemary, lavender—any aromatic herb can clear your head, but those aren't geology hacks.

At SkilledSurvival.com, author Just in Case Jack says: "You have to make split-second decisions, working with what you've got on hand," and a clear mind is essential. You might be familiar with the comment that something has "scared the piss out of him." It's actually a biological reaction that's part of your "fight or flight" wiring. When you get scared, your body pumps adrenaline and other chemicals throughout your system. In many animals, secreting a small amount of urine also occurs. When the nostrils process the ammonia in the air, your senses heighten even further.

In the spirit of bodily functions, consider another readily available "wake-up" hack—the armpit sniff. You may instantly conjure up images from the *Saturday Night Live* skits made famous by actress Molly Shannon. Her character Mary Katherine Gallagher is wont to inhale deeply from her armpit for instant motivation. Healthline.com has a good write-up about the chemical composition of perspiration that justifies Mary Katherine's signature move. Eccrine glands produce most of your body's sweat, which is predominantly water. You also get a whiff of hormones, which can be stimulating. Mixed in are bits of salt, proteins, urea, and the prime ingredient in smelling salts: ammonia.

Urea, or carbamide, is an organic compound of carbon, oxygen, nitrogen, and hydrogen, with chemical formula $CO(NH_2)_2$. The body creates urea to dispose of excess nitrogen when the liver combines two ammonia molecules (NH_3) with a carbon dioxide molecule. In what is known as the urea cycle, highly toxic ammonia converts to urea and gets removed by the body. German chemist Friedrich Wöhler discovered that urea can be produced from inorganic starting materials, startling his fellow chemists of the day. He developed the Wöhler synthesis, showing disbelievers that an organic compound heretofore only known as a by-product of a life process could actually be developed in a lab. For 1828, this was a big deal—the doctrine of vitalism held that only living organisms could produce what were then known as the "chemicals of life." Wöhler thus directly refuted the idea that living things exist only because some "vital force" binds them together.

Biologists have long known that all organisms are built from the same six essential elemental ingredients: carbon, hydrogen, nitrogen, oxygen, phosphorus, and sulfur (CHNOPS). These in turn form the chemicals of life: nucleic acids, proteins, carbohydrates, and lipids, which are the basic chemicals of living things.

When time is of the essence, the original geology hack for reviving yourself consists of smelling salts (ammonium carbonate). Actually, smelling salts should be referred to as an "aromatic compound of ammonia," not a true salt. True salts are halogens; the word "halogen" literally comes from "salt-making." The halogen elements are fluorine (F), chlorine (Cl), bromine (Br), iodine (I), and the little-known astatine (At). All halogens form an acid when combined with hydrogen, but since the compound originally came in a white, powdery form, the salt label has stuck with us.

When a true salt is dissolved in water, it completely breaks down into negatively and positively charged ions. In our case, ammonium carbonate "salt" behaves similarly and breaks down into ions and gasses. For years, athletes used smelling salts as a quick stimulant to clear the cobwebs from a ferocious hit or collision. Writing in the *British Journal of Sports Medicine*, author Paul McCrory notes that the use of

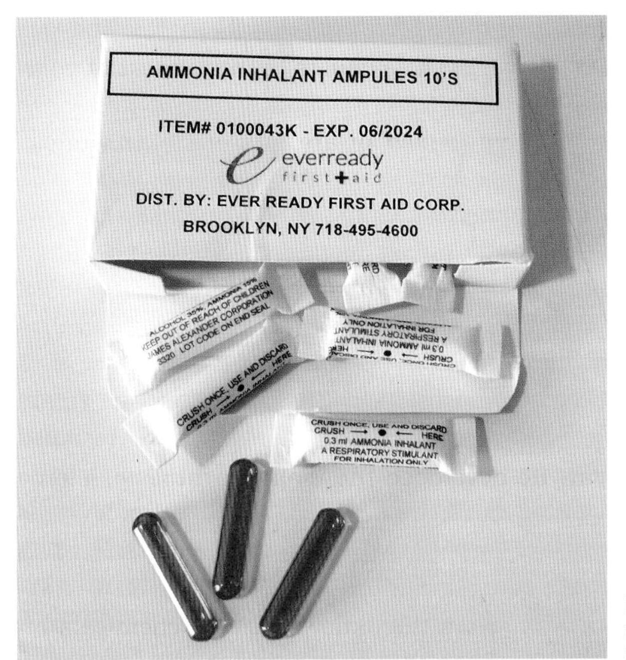

Red ammonia inside a thin glass ampule, ready to crush for instant revival

ammonia-based smelling salts goes back for centuries, to the Romans in 79 CE at least. "The term *Hammoniacus sal* appears in the writings of Pliny," McCrory notes, "although it is not known whether the term is identical to the more modern *sal ammoniac*, which was known to the alchemists as early as the 13th century. Chaucer also noted the existence of *sal ammoniac* alongside a large number of other *materia medica*."

Some first-aid kits contain a sealed glass tube with liquid ammonia inside. After placing the glass in a towel, crushing the glass releases the scent, which you then pass under a patient's nose. Ammonia gas irritates the tender membranes of your eyes, nose, and lungs, triggering a sharp inhalation reflex. This in turn improves respiratory flow rates by squeezing more oxygen into the bloodstream, accompanied by large doses of adrenaline.

At smellingsalts.org, there is more science-based discussion about the biology of inhaling ammonia gas. "When the ammonia gas is inhaled in such strong concentrations," the site states, "it instantly causes some pretty drastic irritation to a number of nerves along your nasal passages: the olfactory, trifacial, and vagus nerves, to be exact. Particularly [important is] the vagus nerve, which is the motor nerve of the heart and bronchi. The vagus nerve is the primary mechanical reason for fainting when exposed to loud noises, lights, and other stimuli. Stimulating this nerve after overload essentially resets the function and wakes up a fainted person."

How to Make Smelling Salts

While smelling salts are inexpensive, there is a way to make the powder at home. Simply add a chunk of dry ice to an ammonia solution and evaporate the water. You'd want to do all of that outside or in a well-ventilated area, or preferably in a chemistry lab under a hood with excellent fans to pull off the stray ammonia fumes.

Of course, dry ice isn't readily available on the homestead, but you're just developing a powder that will give off ammonia gas when mixed back with water. You can save multiple steps by simply bottling a little liquid ammonia so you can take a tiny whiff of the nasty gas and get the same effect.

The Mineral Struvite

According to experts, first-century Roman chemists used a simple method of mixing urine with wood ash and salts then decanting numerous times to purify ammonia. Apparently, "Urine Collector" was a résumé item. In his 2016 article for *Acta Archaeologica*, Michael Witty lays out a plausible scenario whereby Romans would collect urine, use wood ash and magnesium-rich salt to crystallize the ammonium, and form struvite, $MgNH_4PO_4\ 6H_2O$.

As described in *Memoirs and Proceedings of the Chemical Society of London for 1841–48*, German chemist Georg Ludwig Ulex (1811–1883) first described struvite in 1845 after digging up the remains of a medieval midden (a fancy word for a dump) found in the ruins of a church in Hamburg, Germany. He named the new mineral after geographer and geologist Heinrich Christian Gottfried von Struve (1772–1851) of Hamburg. It isn't clear if that was an honor or not, as Ulex noted: "The crystals are the same salt which is found in many animal secretions, and in putrefying urine." He also reported that the source layer was a thick layer of "cattle-dung mixed with straw, in a state of putrefaction" that would also likely revive anyone near it.

At this point in a mythical dire survival situation, you're at least revived. That leaves air, shelter, water, and food. So we'll keep tackling them in that order.

FILTERING AIR

In a true emergency breathing situation, such as being underwater—well, there's no geology hack for instantly concocting breathable air out of water. There are conceptual devices for artificial gills, and you may someday pack such a system in an oceangoing survival kit; but for now, geology isn't much help for underwater scenarios.

If you've fallen under ice, you might be able to locate a trapped air pocket, but chances are, you're better off concentrating your efforts on bashing through the ice with a knee, elbow, or foot. A practical geologist might consider grabbing a rock from the bottom of a shallow water body and hoisting it back up for better ice combat, but think of the time you'd lose trying to find the perfect fist-sized rock.

Most humans can hold their breath for 30 to 60 seconds. In common scenarios, after 3 minutes your brain suffers irreparable damage as vital cells deprived of oxygen begin to die. However, there are exceptions. At GuinnessWorldRecords.com, you can find the story of 56-year-old free diver Budimir Šobat of Croatia. He set a record for voluntarily holding his breath: 24 minutes, 37 seconds. He achieved the feat in December 2020, raising money for earthquake relief.

The most common survival issue regarding breathing is to find good, pure (or nearly pure) air. Thus, constructing some kind of filter is important if you find yourself in a situation such as a severe dust storm or one of the wildfires becoming more common in the western states. You will have much more luck removing sediment from the air than you will extracting dangerous chemical gasses, but both can be done.

Face Mask

Now that the COVID-19 pandemic is fading from top-of-mind consciousness, you doubtless have extra face masks stashed in your car, backpack, pocket, top drawer, etc. At the height of the outbreak, masks were ubiquitous (even in roadside litter). We all became sensitized to the advantages of an N95 mask.

As stated at the FDA.gov website, N95 respirators and surgical masks are examples of personal protective equipment (PPE) designed to protect the wearer from 95 percent of airborne particles and from contaminated liquids reaching the face. "It is important to recognize that the optimal way to prevent airborne transmission is to use a combination of interventions from across the hierarchy of controls, not just PPE alone," the website warns.

An N95 respirator is a respiratory protective device designed to achieve a very close facial fit and provides efficient filtration of airborne particles when working properly. Note that the edges of the respirator are designed to form a seal around the nose and mouth.

At the height of the COVID-19 pandemic, the Centers for Disease Control and Prevention (CDC) and other agencies recommended "double-masking" to further enhance protection. If you found yourself in a blinding dust storm, you'd probably consider triple-masking—double-masking under a handkerchief or some other combination.

Wet Handkerchief

One of the classic survival scenarios is to escape an intense desert sandstorm. Sandstorms are serious emergencies—they ground aircraft, disable vehicles, flatten crops, and kill livestock. And they're getting worse. At ScienceDaily.net, there is a report from University of Kansas researchers who delved into the mechanisms for the massive 2020 dust plume that originated in Saharan Africa and traveled all the way across the Atlantic to the Americas. The storm was nicknamed "Godzilla" due to its ferocity and size. NASA tracked the plume and provided background (svs.gsfc.nasa.gov/4849).

The Sahara Desert occupies 3.6 million square miles across the northern half of Africa, and about 60 million tons of its nutrient-laden sand and soil are lifted into the atmosphere each year. The regular plumes deposit nutrients in the Atlantic Ocean and the Caribbean Sea every year, but the Godzilla storm alone accounted for about 24 million tons of flying debris in four days.

The impact on humans from such a storm can be severe. The Kansas research team pointed to the danger of extremely small particles, with a diameter of less than 2.5 micrometers (0.000098 inch). Such small particles can wedge deeply into lung tissue, causing all kinds of respiratory issues. In addition, the storm cloud "can also contain a dust-dwelling fungus which may cause a disease called Valley Fever that happens in Arizona and California and has been related to dust storms in those regions."

The quick and dirty hack for surviving a dust storm is a wet handkerchief or cloth covering the face. It is only a modestly useful dust and particulate filter, but better than nothing. You have to make sure the fit is tight while still allowing for airflow. The fabric choice is important too—you want to make sure there is some good filtering, and you should double up the fabric. You may not have access to water, but wetting the fabric can assist greatly, and might even be worth a few drops from your canteen.

Add Baking Soda

Wetting your fabric filter with a baking soda solution might temporarily block some gasses that are toxic. Baking soda, or sodium bicarbonate, has the formula $NaHCO_3$. It is typically present as a salt, composed of a sodium cation (Na^+) and a bicarbonate anion (HCO_3^-). The natural mineral form is nahcolite, which is found in mineral springs.

Since baking soda is produced on a massive scale, it is usually quite cheap to purchase. Still, there is a way to make your own.

1. Dissolve lye in water.
2. Let sit for a long period (days to weeks) to absorb carbon dioxide from the air.
3. Slowly evaporate the excess water (at low temperatures) then collect the residue.

If you don't have lye, there are multiple recipes available on the internet. Basically, you do the following:

1. Boil the ashes from a hardwood fire for about 30 minutes (softwoods are too resinous), and use rainwater if possible, as it is "soft"—rain water lacks extensive mineral components of spring or creek water.
2. Allow the ashes to settle to the bottom of the pan.
3. Skim the liquid lye off the top. Use caution; lye is extremely caustic and causes serious chemical burns.

Lye is a metal hydroxide, usually based on sodium, with a chemical formula NaOH. Another form is a potassium hydroxide, KOH. Pioneers and homesteaders typically produced lye by leaching hardwood ashes. They then mixed lye with water and fat to create soap.

HealthLine.com has a good description of twenty-two benefits and uses for baking soda. The uses range from teeth or laundry whitener to workout enhancer and room deodorizer. Most kitchen odors, for example, are acidic in nature, so basic baking soda neutralizes them through chemistry.

While there isn't a ton of research on the long-term health effects of wildfire smoke, many experts believe that inhaling particles from wood fires can affect the immune system and potentially trigger asthma, allergies, and breathing problems. It's something to worry about, as wildfires are increasing in scope and frequency. Fortunately, there are products available to help you.

One of the survival products at SmokeyZone.com is a high-end handkerchief, available in black, camo, and green. At $57 it's a bit pricey, but it can assist in fire-crafting, purifying water, filtering toxic air, and more. It's the "multi-tool" of rags, based on a 1-micron nominal nonwoven filter. The manufacturer claims it can treat water on the fly by serving as a straining system, or serve as a hand-to-mouth mask to shield from the effects of gases, smoke, and other airborne contaminates. It's a nonwoven food-grade polymer fiber that dries fast to inhibit mold as well. It can pass 1 cubic foot per minute (CFM) of air and is claimed to be effective against carcinogens, exhaust, and other pollutants (such as formaldehyde, toluene, hydrogen sulfide, and ammonia). It also serves as a barrier against harmful gasses such as benzene and carbinol and airborne pathogens.

Plastic Bottle Air Filter

One ingenious hack from TotoBobo on YouTube is to improvise T-shirts (or cloth) and a plastic water bottle to create a filter. The shirts should be free from scented soap, dirt, oil, etc.

1. Remove the cap from a clean plastic water bottle, but stay clear of the larger 2-liter size.
2. Cut four or five holes in the bottle spaced out, varying in size from penny, nickel, or quarter—¼ to ½ inch in size.
3. Fold a T-shirt into a long slab of cloth matching the length of the bottle.
4. ***Optional:*** Add a layer of very fine sand, ground up charcoal, dried mint leaves, etc., between the layers when folding up the cloth.
5. Wrap the cloth around the bottle, allowing access to the top.
6. Repeat one or two times, if you have enough spare T-shirts.
7. Tie the wraps tightly in place.

You can now breathe in from the top of the bottle, drawing air through the cloth and into the chamber. Using several layers of cloth dramatically reduces the particulates entering your lungs. There are many variations on this type of filter. You could add (clean) cigarette filters to the mouth opening, for example, or substitute cattail fibers.

You could also partially fill the bottle with coarse sand, cattails, leaves . . . the variations are endless. You'd have to experiment with all the variables to make sure enough clean air comes through the filter. You might experiment with a hookah design, as well.

Activated Charcoal

There is quite a bit of science behind charcoal. A typical outdoor wood fire leaves behind partially-burned wood, or charcoal, and it's a good resource on its own. It has qualities that make it ideal for filtering and purifying both air and water. Activated charcoal is like super-charcoal; it concentrates all the good qualities of regular charcoal.

At MomWithAPrep.com, author Sean Kerr describes the advantages of activated charcoal as based on a larger surface area of exposed carbon. Activated charcoal has exponentially more nooks and crannies where carbon atoms are exposed. Emergency room doctors use activated charcoal taken as capsules or a slurry to assist with overdoses; the charcoal absorbs harmful toxins from the stomach contents before they can enter the bloodstream. If you have indigestion due to gas, activated charcoal capsules help neutralize the gas.

It may seem counterintuitive, but the black activated charcoal powder can whiten teeth and helps with skin care too. It assists with composting, treats rashes, and is a good passive air purifier—small pouches hung around a room can absorb odors.

The recipe to create your own activated charcoal is fairly simple.

1. Pack pieces of dry hardwood in a metal pot. The pot should have a hole for ventilation so you can monitor smoke production.
2. Set the pot on a modest fire and burn for at least 3 hours, or until there is no more smoke or gas escaping through the vent hole.
3. Remove from heat and allow to cool.
4. Rinse the charcoal in clear, cool water. This removes debris and ash.
5. Dry for several hours, at least. The more water you can evaporate from the charcoal, the better.
6. Grind the dried charcoal into a very, very fine powder using a mortar and pestle. Or you can place the charcoal in a bag and smash it into a powder with a mallet.
7. Allow to dry for several more hours. You will probably have exposed damper charcoal in the crushing process; set it aside.

8. Add a mixture of 1 part calcium chloride to 3 parts water to the powdered charcoal in a stainless steel bowl so that you cover the powdered charcoal completely and create a thick paste. There are alternatives to calcium chloride—use the same ratio for lemon juice or bleach, for example.
9. Let the mixture steep for at least a day.
10. Drain off most of the excess liquid.
11. Bring the mixture to a boil in a clean metal pot and boil for at least 3 hours. This is the step that activates the charcoal.
12. Remove from heat, cool, and evaporate any remaining moisture.

Create Your Own Calcium Chloride

If you don't have bleach or lemon juice and want to make your own calcium chloride ($CaCl_2$) from limestone ($CaCO_3$), you'll need hydrochloric acid (HCl).

Here's the process:

$$CaCO_3 + 2HCl \rightarrow CaCl_2 + CO_2 + H_2O$$

Calcium carbonate plus hydrochloric acid yields calcium chloride plus carbon dioxide plus water.

1. Fill a glass beaker or jar about one-quarter full of clean, dry limestone gravel.
2. Add an equivalent amount of hydrochloric acid.
3. Allow to bubble completely, avoiding the CO_2 released. Stir as necessary, avoiding acid spills. If you completely dissolve the limestone, add more of it.
4. Pour off the resulting liquid through filter paper into another glass container.
5. Heat the solution gently until dry to drive off the water molecules. When complete, you should see small lumps of pure white calcium chloride crystals.

Create Your Own Hydrochloric Acid

If you don't have hydrochloric acid available, there's a recipe for that too.

Use a ratio of 1 gram of salt to 60 milliliters of vinegar.

$$HC_2H_3O_2 + NaCl \longrightarrow HCl + NaCH_3O_2$$

Vinegar + salt —> hydrochloric acid + sodium acetate

Pour off the acid solution (carefully) and dry the sodium acetate gunk. Primarily used as a desiccant to remove moisture, sodium acetate is also used to neutralize sulfuric acid in the textile industry, as a pickling agent in tanning, as a food seasoning to replicate salt and vinegar, and as a concrete sealant.

Hand warmers make use of sodium acetate by taking advantage of its low melting point: around 137 F (58 C). When you heat the crystals past the melting point and then cool, the solution becomes supersaturated. It will not form crystals at room temperature. By pressing on a metal disc within the warmer, you create a seed crystal or nucleation center. This initiates the crystallization process, which chemists call "exothermic"—it gives off heat. Unlike some "one-way" heat packs, a sodium acetate pack can go back and forth. Immerse it in boiling water to remelt the crystals, allow the pack to cool to room temperature, and it's ready to go again.

If you need to go even further down the manufacturing process, you can collect salt by boiling off seawater. If you need to create your own vinegar, there are countless methods available, but this is leaving the realm of practical geology and ranging into practical chemistry. Some vinegar recipes rely on purchasing yeast; others use a SCOBY (symbiotic culture of bacteria and yeast), sometimes referred to as "the mother" because it can live on and on, as long as you feed it, and can produce vinegar, kombucha tea, and other products.

Filtering Indoor Air

If you simply want to purify indoor air, rather than survive an immediate emergency, there are numerous tips and tricks available. At Welland Good.com, there is a discussion about steps you can take, ranging from opening the windows to growing indoor plants.

Researchers for the *Journal of Exposure Science and Environmental Epidemiology* found that it would take between 10 and 1,000 plants per square meter of floor space in a house to compete with the air-cleaning power of a couple of open windows.

Ultraviolet (UV) Lights

To clean both air and water, some manufacturers have developed special UV lights that kill bacteria. These lights can even help sanitize a home when an occupant suffers from a cold or another illness.

A 2020 article at HealthLine.com goes into depth concerning the ability of UV light to kill bugs and germs. But first, it's important to understand just which part of the UV spectrum is most reliable.

- **UVA light** has the lowest amount of energy, and makes up most of the light you get from the sun. Although low-energy, UVA has still been linked to skin damage.
- **UVB light** is in the middle of the UV light energy range, and makes up only a small portion of sunlight. It causes sunburn and most skin cancers.
- **UVC light** has the most energy, and causes the most damage, but is greatly absorbed by our planet's ozone layer.

In a sidebar, the HealthLine article describes a potential future tool: a room-clearing robot using UV light reportedly can kill 99.99 percent of SARS-CoV-2 (COVID) viral particles in 2 minutes. It's possible that robots like this could disinfect hospital rooms, hotel rooms, and airplanes in the future. However, there are trade-offs—UVC light can fade textiles and plastics, damage your eyes and skin, and could have other side effects.

CHAPTER 4

SHELTER FROM THE STORM

Once you are alert and breathing good air, you must seek shelter. Water can come later; you will perish from exposure to the elements far faster than succumbing to thirst.

Dying from exposure is due to a lack of protection from extreme temperatures, environmental conditions, or dangerous substances. Exposing the human body to the elements, such as in a snowstorm or desert heat, leads to exposure—a catchall phrase with multiple meanings. In most cases, you are losing the race to keep your body comfortable.

CHOOSING A SPOT

In a desperate survival situation, you may not have time or strength to search carefully for the best place to seek shelter. Author Hugh McManners describes some of the issues and solutions in his epic book *The Complete Wilderness Training Manual* (Dorling Kindersley, 1994). He lists the challenges, beginning with safety, warning: "You may not be able to identify all the potential hazards or disadvantages of a particular spot, but in order to give yourself the best chance, you should allow plenty of time before dusk to look for a campsite" (McManners, 40).

McManners suggests avoiding the very bottom of a valley, as water chases gravity; he also warns you to consider avalanche potential, flash flood risks, and potential for falling rocks and trees, as well as prevailing winds. Even the hollowed-out trunk of a large tree will work.

This hollow cedar tree in the Olympic rain forest would make a decent survival shelter in a rainstorm—and likely has.

McManners has definite views on "an ideal campsite." His advice is summarized in table 14, with the addition of seeking open skies.

Table 14: Campsite Considerations When You Have Enough Time to Choose a Good Location

Issue	Resolution
Prevailing wind	Detect the major air currents to generally avoid smoke from your fire and cut down on latrine odors.
Sheltering trees	Avoid snags that can blow over in strong wind. Be close to a wood supply. Set up your campfire safely so as to avoid overhanging branches.
Accessing the water	Choose access to get fresh water upstream. Dump waste downstream. Avoid getting too close to water, as flooding, cave-ins, bogs, and swamps are all dangers.
Viewing the sky	Find a locale with some kind of view in order to study the night sky, gain southern exposure, assess long-distance safety issues, and plot next steps for when you eventually move.

In wetter climates, you'd do well to dig rain channels away from your tent or shelter. This captures the rainwater and drains it away so that you don't wake up to a damp floor. Use a digging stick or a strong, flat rock if you don't have a shovel.

SURVIVING THE COLD

Let's tackle freezing to death first, otherwise known as hypothermia. The Mayo Clinic has a simple definition for hypothermia: "A medical emergency that occurs when your body loses heat faster than it can produce heat." Most of us know that the human body normally operates at around 98.6°F (37°C). Hypothermia occurs as your body temperature falls below 95°F (35°C). As the body's core temperature falls, the heart, nervous system, and other organs stop functioning properly. The treatment is to warm the body back up, either through heat wraps, extra layers, or—in dire emergencies—using hot water immersion.

At LiveScience.com, senior writer Rachel Rettner describes the conditions for hypothermia, and notes that some people "can experience hypothermia in relatively cool, but not freezing air temperatures—around 30°F to 50°F (-1°C to 10°C)—particularly if they are wet, such as from rain, sweat, or submersion in cold water." She quoted Michael Sawka, chief of the Thermal and Mountain Medicine Division at the US Army Research Institute of Environmental Medicine (USARIEM), who

Survival in extreme conditions requires planning, patience, and experience.

told *Live Science* in a 2010 interview that the body loses heat twenty-five times faster when wet.

AmericanOutdoor.guide has a lengthy section on winter survival skills. Author Larry Schwartz goes into good detail about how to "beat Old Man Winter" at his own game. Schwartz boils down the rules in an acronym: C.O.L.D.:

C Keep CLEAN—in body as well as clothing.
O Avoid OVEREXERTION—you can build up sweat, wear out too fast, and make bad decisions.
L Use LAYERS to control your temperature.
D Stay DRY to maximize warmth.

He describes the different ways the body loses heat:

Evaporation—When sweat evaporates from your body, it takes heat with it.

Radiation—When your body temperature is higher than the air (or water) around it, the body will radiate its heat away. The greater the difference between body temperature and surrounding temperature, the more heat is lost.

Conduction—Heat is conducted away by contact with air and water. Air is a poor conductor compared to water, and fluffy material serves as insulation. As anyone lying on cold earth or a rock knows, such materials tend to conduct heat quickly away from the body.

Convection—Moving air or water continually dissipates heat by supercharging the conduction effect. A common example of this is a breeze or wind blowing across your body, causing windchill.

Expiration—Each time you breathe out warm, damp air, you lose heat in the form of warm water vapor.

Temperature (°F)

Wind (mph) / Calm	40	35	30	25	20	15	10	5	0	-5	-10	-15	-20	-25	-30	-35	-40	-45
5	36	31	25	19	13	7	1	-5	-11	-16	-22	-28	-34	-40	-46	-52	-57	-63
10	34	27	21	15	9	3	-4	-10	-16	-22	-28	-35	-41	-47	-53	-59	-66	-72
15	32	25	19	13	6	0	-7	-13	-19	-26	-32	-39	-45	-51	-58	-64	-71	-77
20	30	24	17	11	4	-2	-9	-15	-22	-29	-35	-42	-48	-55	-61	-68	-74	-81
25	29	23	16	9	3	-4	-11	-17	-24	-31	-37	-44	-51	-58	-64	-71	-78	-84
30	28	22	15	8	1	-5	-12	-19	-26	-33	-39	-46	-53	-60	-67	-73	-80	-87
35	28	21	14	7	0	-7	-14	-21	-27	-34	-41	-48	-55	-62	-69	-76	-82	-89
40	27	20	13	6	-1	-8	-15	-22	-29	-36	-43	-50	-57	-64	-71	-78	-84	-91
45	26	19	12	5	-2	-9	-16	-23	-30	-37	-44	-51	-58	-65	-72	-79	-86	-93
50	26	19	12	4	13	-10	-17	-24	-31	-38	-45	-52	-60	-67	-74	-81	-88	-95
55	25	18	11	4	-3	-11	-18	-25	-32	-39	-46	-54	-61	-68	-75	-82	-89	-97
60	25	17	10	3	-4	-11	-19	-26	-33	-40	-48	-55	-62	-69	-76	-84	-91	-98

Frostbite Times: 30 minutes, 10 minutes, 5 minutes

$$\text{Wind Chill (°F)} = 35.74 + 0.6215T - 35.75(V^{0.16}) + 0.4275T(V^{0.16})$$

Where, T=Air Temperature (°F) V=Wind Speed (mph)

Effective 11/01/01

Wind Chill Calculation Chart WEATHER.GOV

When you combine wind speed and low temperature, you get wind-chill. For example, as shown in the wind chill chart, even a 40°F day with a wind speed of 40 miles per hour can bring on frostbite in 27 minutes.

The immediate goal when facing hypothermia is to bundle up and get protected from the elements so that you build a shelter. Fierce winds can drive heat from your skin through conduction and convection almost as quickly as water can. You'll want to protect every bit of skin from exposure, and there are volumes of material available discussing the best clothing for survival situations. The first rule is to add layers. There are base layers, inner layers, outer layers . . . the list is endless.

One of the cruelest stages of hypothermia is called "paradoxical undressing." It is often reported from fatal hypothermia cases, where victims are suddenly overcome by a feeling of warmth, if not extreme heat. Despite low environmental temperatures, victims paradoxically remove their clothing, putting themselves at even greater risk. As described in a 2013 article by Marc Lallanilla at LiveScience.com, to shut down the loss of heat from the extremities, "the . . . muscles necessary for inducing vasoconstriction become exhausted and fail, causing warm blood to rush from the core to the extremities. This results in a kind of 'hot flash' that makes victims of severe hypothermia—who are already confused and disoriented—feel as though they're burning up, so they remove their clothes."

Another odd effect of late-stage hypothermia is called "terminal burrowing." According to the same article, victims in the final throes of severe hypothermia often exhibit a bizarre behavior similar to beginning hibernation. Referencing a 1995 article in the *International Journal of Legal Medicine*, Lallanilla cites researchers from Germany who described hypothermia victims "in a position which indicated a final mechanism of protection, i.e., under a bed, behind a wardrobe, in a shelf, etc."

One key tactic agreed upon by survival experts is to avoid rash decisions. There is another common acronym for surviving extreme conditions:

S **Slow** down to a stop.
T **Think** about what you are doing.
O **Observe** the terrain and environment.
P **Plan** out what you need first, second and afterwards.

Next comes the feat of building some kind of structure. There are countless resources available to show you how to fashion a quick structure with a tarp and rope, for example. If you don't have supplies, you'll need to get an assist from geology. Consider some of the following options:

1. **Find an overhang.** Cliffs and outcrops often make good temporary camping locations. You can pile up rocks at the ends to minimize the wind, or bank up soil.

Small cliffs and outcrops offer an easy opportunity for a quick survival shelter.

Natural caves offer fast protection from extreme heat and extreme cold.

2. **Locate a cave.** Caves are the original shelter hack, but there are all kinds of caveats—something may already be living there, running water may cause issues, whipping wind may rush through and cause more heat loss. The darkness can conceal hidden dangers such as loose rock, sharp drop-offs, and other sources of mechanical injury. But to just get out of the elements and find protection, caves are ideal.
3. **Pile up debris.** The classic survival hack is a debris shelter or hut, outlined something like a small pup tent using wooden poles with forks on top, joined by a long pole, then heaped with insulating leaves, soil, moss, and/or branches. You can also use an A-frame technique or go tepee style like North American Indians. You especially want to minimize heat loss by adding insulation between you and the floor, keeping the shelter small, and checking the terrain for likely water flow in heavy rain.

You can combine rocks and wood for a fast emergency shelter, but debris huts are your best option.

Fallen logs provide a fast shelter opportunity as well as firewood.

4. **Combine rocks and wood.** If you locate a jumble of large boulders, you can use them for walls and cover the top with branches. Then use the debris-hut methodology for insulation.
5. **Build an ice cave or igloo.** If you have a saw and find yourself in a snowstorm, you can build a structure out of ice and snow, gently sloping the walls to a round dome. Pack the crevices tightly with loose snow, smoothing as necessary. Again, consider insulating the floor.

At PreppingToSurvive.com, the authors emphasize practical geology for improving your cold weather survival shelter. "Make use of naturally occurring formations," the experts suggest. "The downwind side of a downed tree, the side of a relatively large dirt mound, or the side of a large rock all make for a good start to a shelter. Add some support sticks, leaves, and boughs."

Heat loss through the cold, damp earth is one of the primary issues to address. Your temperature will fade rapidly through conduction as the ground greedily gobbles up your excess body heat. Use heaps of dry foliage, such as leaves, needles, or tree boughs, as an easy insulation source. If water removes body heat twenty-five to thirty times faster than air, cold damp soil is nearly as efficient at conducting heat away. Try to get dry.

It takes time, effort, and planning to build your first survival structure, but the key is to be methodical, deliberate, and efficient. It takes practice, too—your second shelter will probably look better than your first. If you plan to stay hunkered down for a long stay, you'll need to consider how to keep a fire burning in the structure by building a fireplace. Larger structures like a tepee do well in allowing smoke to rise, but if you're up against a stone wall, you can fashion rocks as a makeshift fireplace and chimney.

SURVIVING THE HEAT

When your body temperature rises above a certain level, the result is hyperthermia—heatstroke. In hot, humid climates with little or no access to water or shade, you'll perish just as surely as you would if caught in a snowstorm.

At OutsideOnline.com, Amy Ragsdale and Peter Stark compiled an excellent resource about heatstroke. They explain that more US citizens die annually from heat-related illnesses than from hurricanes, lightning, earthquakes, tornadoes, or floods. There were more than 9,000 heat-related deaths in the United States between 1979 and 2014. Some

Even a small amount of shade can make a big difference, as well as provide a place and time to stop and think about next steps.

years, the numbers are far worse; in 2021 a huge heat dome formed over the usually temperate Pacific Northwest states and western Canada, resulting in temperatures up to 20°F higher than normal and causing about 700 deaths, according to estimates from the Canadian Broadcasting Corporation and other outlets.

In August 2003 in Europe, according to a report recapped at New Scientist.com, at least 35,000 people perished due to record heat that reached over 100°F and lasted for several days. Experts predict that global temperature increases and extreme weather events will accelerate if greenhouse gasses continue to pour into the Earth's atmosphere. Geology may have a temporary answer for that too: A phenomenally huge volcanic eruption would slow down rising temperatures. In April 1815 Mount Tambora erupted in Indonesia; the ash cloud quickly circled the globe and hung in the atmosphere so long, 1815 was dubbed "The Year without a Summer." An erupting caldera or super-volcano would do even more damage; check the book *Super Volcanoes* by Robin George Andrews (W. W. Norton & Co., 2022) for more insight.

The human body suffers much more from extreme heat than extreme cold. According to Ragsdale and Stark, the lowest body temperature a human has been known to survive is 56.7°F, nearly 42 degrees below normal. "Anna Bågenholm, a 29-year-old Swedish woman, was backcountry skiing when she broke through 8 inches of ice into a frozen stream. Her upper body was sucked down, leaving only her feet and skis visible, but she managed to find an air pocket and was able to breathe. She was finally rescued after 80 minutes. Bågenholm remained in a coma for about ten days and was in intensive care for two months but ultimately suffered only minor nerve damage."

A swing of 42 degrees downward is a major event, but it takes much less to create havoc from heat. The same authors report that Willie Jones, a 52-year-old Atlanta man, endured an increase of only 17 degrees above normal when his body temperature reached 115.7°F in a 1980 heat wave. Jones barely survived, and was finally released from the hospital after twenty-four days.

Fortunately, the same rules about insulation, convection, and radiation in an extremely cold situation apply to survival situations involving excess heat. You want your structure to emphasize roof and wall insulation, especially on top. In fighting heat, however, you might want to keep the earthen floor cleared to allow for the cooler earth to keep your body temperature regulated. You'd want to allow wind currents

Caves offer a shady spot during the height of the day's heat, but use caution. If bats are present, leave them alone.

to sweep through the structure, and you'd definitely want to be near a water source as well as shade.

Rocks under direct sunlight would be a less desirable construction material in severe heat, as they tend to absorb radiation and release it slowly. But with the wild temperature swings known in desert climates, where temperatures from day to night can fluctuate by 40 degrees or more, that nighttime heat source might be a good thing. As with all survival situations, your mileage will most certainly vary.

SURVIVING DUST STORMS

Dust storms and sandstorms are among nature's most violent and unpredictable phenomena. High winds lift dirt or sand particles into the air, unleashing a turbulent, suffocating cloud that can reduce visibility to almost nothing in a matter of seconds and cause property damage, injuries, and deaths. No matter where you live, it's a good idea to know what to do if you see a wall of sand racing toward you.

The most celebrated dust storm in US history was called "Black Sunday" by its survivors, and occurred in the area that was to become known as the "Dust Bowl." Beginning on April 14, 1935, and lasting for days, it was one of the worst dust storms in American history, covering Texas, Oklahoma, Kansas, Colorado, and New Mexico. Scientists estimate that the storm stripped perhaps 300 million tons of topsoil from the prairie.

Lawrence Svobida, a Kansas wheat farmer during the 1930s, wrote in his memoir, *Farming the Dust Bowl*, that the approach of yet another dust storm was a sight to behold.

Haboob bearing down on Phoenix, Arizona, in 2018 RACHEL HOUGHTON.

> A cloud is seen to be approaching from a distance of many miles. Already it has the banked appearance of a cumulus cloud, but it is black instead of white and it hangs low, seeming to hug the earth. Instead of being slow to change its form, it appears to be rolling on itself from the crest downward. As it sweeps onward, the landscape is progressively blotted out. Birds fly in terror before the storm, and only those that are strong of wing may escape. The smaller birds fly until they are exhausted, then fall to the ground, to share the fate of the thousands of jackrabbits which perish from suffocation.

As discussed earlier, first, cover your nose and mouth with a mask or damp cloth to reduce exposure to dust particles; double up or even use three layers if possible. Finally, wrap a bandanna or some other piece of cloth around your nose and mouth, over the top of your face mask if possible. A shirt sleeve or a medium-sized sheet works only as an emergency guard, as it leaves gaps over your face. Moisten it a bit if you have enough water.

What then? There's a good write-up at wikiHow for surviving a dust storm or sandstorm. Below is a summary of tips and tricks based on the discussion:

1. **Prepare for dryness.** Apply a small amount of petroleum jelly to the inside of your nostrils to prevent drying of your mucus membranes. Cooking oils or even animal fats work in a pinch.
2. **Add eye protection.** Airtight goggles are far superior to sunglasses or regular eyeglasses. Guarding against flying debris, ranging from small particles of dust to larger pebbles and sticks, is important. Even holding your arm up over your face works in an emergency.
3. **Use eye drops.** Specialty eye drops contain lubricants and soothing medications such as polyethylene glycol, polyvinyl alcohol, propylene glycol, glycerin, and mineral oil. Dry eyes lead to a variety of issues, from headaches to eye irritation. You can wash your face with clean water, but you don't want to introduce anything into your eyes that can cause infections later on.
4. **Look for shelter.** You may have to simply wait out the worst of the storm in some form of leeward (downwind) protected area. Large rocks, parked cars, tree trunks, walls, hedges, shrubs—anything to protect your eyes and nose is crucial. If caught out in the flat

This fire pit could also double as an emergency shelter.

desert, you may have to sit down with your back to the wind, taking care to mark how much debris is building up around you.

5. **Get to high ground.** This may seem counterintuitive, but just like a gold-bearing stream carries the heavy gold in the bottom of the current, a windstorm concentrates the heaviest material near the bottom of the cloud. If you can get to higher ground, you'll avoid the worst of the destruction. Just make sure you don't end up in a structure that actually funnels the wind and makes things worse.

There are a few more geological considerations to keep in mind if you find yourself on foot in a dust storm:

1. **Avoid lightning attractors and exposed areas.** Dust storms often include lightning storms, which can seek out large trees or exposed humans. You'll know if you are in such a situation from the booming thunder and lightning displays.
2. **Prepare for flash flooding.** Many dust storms also include rain showers of such intensity that ditches, creek beds, gullies, and gulches can turn dangerous. Such channels may offer temporary respite from the flying projectiles, only to flood out in a rumbling, crashing flash flood. If you're in a gulch and can look around, check for water-carried debris in the branches above your head, or look for telltale water lines on boulders or cliffs. Rain storms 20 miles away can result in a flash flood under your feet.
3. **Ride it out.** If riding a bicycle, ATV, 4WD vehicle, horse, or camel in the desert, consider hunkering down in place, with the mode of transportation protecting you from the wind. Camels easily survive dust storms, calmly waiting things out. Horses are more finicky, but also see the wisdom in waiting out the storm rather than walking in it.
4. **Understand fluid dynamics.** Be cautious on the leeward side of a sand dune in a storm, as that's where sand deposits build up first. Sand dunes continually move, grow, recede, and react to a storm. The leeward side of a dune could easily become a sand trap.
5. **Stick together.** Stay with the group, and don't venture out until the danger has passed. You will make far better time once the danger abates, and the group could end up wasting valuable energy trying to track down wayward souls who stubbornly venture out on their own.

6. **Conserve water.** We'll talk more about this later, but it's never too early to begin conserving water in an emergency. If you are stuck in a dry dust storm, you could face the prospects of minimal water resources due to piled up sand and debris.

MANAGING MUD

One of the oldest geology-based survival hacks employed by humans is applying mud to the skin to ward off bugs and prevent sunburn. Any wet soil will work, because just about any sample contains enough clay to coat and stick. A layer of wet mud applied quickly can ward off ultraviolet rays sufficiently to prevent blistering and redness. Insects will be discouraged from biting if you have a caked layer between them and your skin.

It's likely that ancient survivalists took their cues from the animals they observed. Elephants, hippos, rhinoceroses, Cape buffalo, and warthogs are some of the many mammals with a known penchant for muddy spa treatments to protect their hides.

The famed "Great White Ghosts" of Namibia's Etosha National Park are elephants that wallow in a calcite-rich mud that dries to a ghostly white. The Sumatran rhinoceros can spend up to 3 hours at a time wallowing before foraging, while captive rhinos that can't wallow show all kinds of skin conditions, including inflammations, hair loss, and suppurations. Animals gain multiple benefits from wallowing in mud—tiny ectoparasites vanish from American bison, while warthogs gain camouflage. Some animals even employ dust if wet mud isn't available.

In extremely sunny climates, a daub of dark mud under the eyes, similar to the black eye grease used by athletes, can reduce glare from pale skin and helps contrast light and dark objects.

Naturally, there are always risks. You may be unlucky enough to be surrounded by strongly alkali deposits that can cause a chemical burn in tender areas. You might accidentally bring coarse sand in with your sample and scratch yourself. Drying mud can itch, and when you sweat, you might end up with mud in your eyes. You would hate to have to use precious water to make mud in a desert climate, and substituting urine only makes you smell worse. Still, if the sun damages your skin until it blisters, your odds of surviving extreme heat plummet. So in extreme cases, a little smelly mud is better than nothing.

FINDING YOUR WAY

Geology and astronomy offer a pair of survival hacks to keep going in the right direction as you seek shelter. Survival 101 dictates that you follow watercourses downstream, on the almost always valid idea that cities and towns eventually await as the stream joins a river. Barring the choice of following water, or in addition to it, you'll need options for guessing the compass points and determining the time of day. Most survival kits feature a compass, but if you don't have one, there are options.

Simple Sundial

One of the oldest navigation methods is the simple sundial:

1. Push a 3-foot stick into the ground so that you can see a shadow. (Obviously this hack won't work on a cloudy day.)
2. Mark the location of the end of the shadow with a small pebble.
3. Wait 15 minutes and mark again with another pebble.
4. With your back to the stick, place your left foot on the first pebble and your right foot on the second pebble. You will face north.
5. Draw an imaginary line from the second rock marker to the first. It should point west.

Primitive Clock

It's vital that you leave enough time in the day to seek out a good place to hunker down. If you have a full view of the southern horizon, use this method to crudely guess the time of day:

1. Determine due east and lay out a line of pebbles to mark it. This should correspond to where the sun came up.
2. Rotate 180 degrees and mark out due west. This corresponds to where the sun will set.
3. Divide the interval into twelve sections, which will roughly correspond to the 12 hours between 6 a.m. and 6 p.m. At the summer equinox, June 21, the sun is above the horizon from roughly 6 a.m. to 10 p.m., while at the winter equinox, the sun is restricted to 8 a.m. to 5 pm, so adjust accordingly.
4. Note the position of the sun against your marks.

Survival Compass

A classic example of practical geology is the creation of a survival compass. You'll need a magnet, a needle, and cork, but you can substitute a dry leaf in calm water. If you don't have a magnet, there is a hack for that too. First, the example with a magnet:

1. Fill a shallow glass bowl, plastic pan, or ceramic dish to at least halfway with water and allow to settle completely. You can get the same results with a puddle, but make sure there is no flow, such as in a stream.
2. Place a small flat disk of cork, a piece of leaf, or even a piece of thick paper in the center of the pool of water.
3. Rub a magnet along the needle in one direction only. If you have a strong neodymium magnet, you may not need to rub the needle more than five times. If you're using a weaker refrigerator magnet, you may need to stroke the needle as many as fifty times.
4. Place the needle on the cork or leaf. Some experiments advise piercing the cork with the needle or taping the needle to the paper.
5. Allow the needle to settle. It should be aligned at the north and south magnetic poles.

There are variations to this trick. If you know the north end and south end of the magnet, you can rub one end of the needle with the north pole and the other end with the south pole. You would want to mark the north end of the needle if possible, but this would have the advantage of instantly identifying north. The downside is that it usually requires a compass to identify the poles on a magnet, as the south pole of the magnet will attract the compass.

If your magnet doesn't weigh too much, you can create a compass by floating the magnet on a piece of wood, Styrofoam, or a leaf. The lightweight, floating platform gives the magnet a nearly frictionless surface upon which to freely rotate. The north pole quickly becomes obvious.

If you don't have a needle, you can substitute a slender strand of steel wire. Copper won't work.

If you don't have a magnet, you can take the needle from your survival kit and rub it on silk or some other electrostatic fabric, such as the sleeping bag interior that creates sparks when you move at night. That should be enough to generate a weak magnetic charge.

GEOLOGIC SHELTER HACKS

Geology furnishes survivalists with a wide variety of bush craft hacks and shortcuts for avoiding the elements. Two of the key hacks are fairly obvious: hot springs and caves.

Hot Springs

One way to warm up fast is to find an active hot springs. The spring's hot water can stave off hypothermia, and after the springs flows a ways, it usually will be cool enough and suitable for drinking, although it may have a bit of an odor if it is particularly sulfurous. There are numerous guides to hot springs available, and HotSpringsofAmerica.com has a locator. Keep in mind that hot springs attract people, so if you're trying to overcome the sudden onset of a nuclear winter, keep security in mind as you approach.

For a reliable heat source, nothing beats a soak in a hot springs. In most cases the water is safe to drink, if a bit stinky.

Caves

Historically, caves have often featured prominently in survival scenarios, and for good reason. They offer an immediate strong roof and thick walls for safety and protection from wind, snow, and rain. The more modern approach to letting geology do the work for you is to consider abandoned mines as a place for temporary shelter. Let's talk about both.

On the basis of the fantastic artwork painted on the walls and recovered tools and artifacts, we know that ancient humans survived in part by locating in caves throughout South America, Africa, and Europe. However, the stereotype of a "cave man" is mostly wrong; just a tiny fraction of humans ever lived in caves.

Derek Ager, emeritus professor of geology at the University of Swansea in Wales, wrote at NewScientist.com in 1992: "Natural caves and rock-shelters are largely restricted to limestone country . . . [however] vast areas of the land surface, on which early humans must have lived, simply have no caves." He further pointed out that countless stone axes were left behind by ancient inhabitants of Kenya, but there were no caves. "I saw no caves in the Olduvai Gorge in Tanzania, where early men lived for so long, and . . . no caves along the shore of Lake Turkana in Kenya, with its famous hominoid fossils."

Consider the risks to cave dwelling—bears and other predators may already live there, so you'd have to drive them out first. Plus the interiors are usually dark and rocky, with uneven floors and ceilings, inviting the possibility of both drafty conditions and lack of airflow. The presence of bats, and attendant bat guano, would have produced an atrocious

The sandstone hills at Olduvai Gorge yield ancient human fossils—and have zero caves. JOE CALHOUN

stench. Limestone caves are notorious for running water percolating through the cracks, at least at the lower levels.

Ager agrees that caves were important to some of our ancestors, as well as useful to paleontologists looking for evidence of past life. As he notes, the first Australopithecinae were found in caves in South Africa. The so-called "first European" was found in the Petralona Cave in Greece, while early traces of *Homo sapiens* came from Kent's Cavern in the Devonian limestone of south Devon and the caves in Carboniferous period limestone in Wales. He thinks it more likely that caves functioned primarily as holy places. "Whatever they were, they were probably not the homes to which the cartoon image of cave men regularly dragged cave women by their hair."

Thus, rather than survival settings, caves offered a different resource—a place for vision quests, religious rituals, and spirit conjuring, and a canvas for artwork that would survive longer due to less exposure to the elements.

Survivalists have for years known that caves are the ultimate hack for fast shelter. The openings of even the smallest cave systems provide relief from snow, rain, and howling winds. Larger systems can house

The Toquima Cave in Nevada was likely used by shamans for spirit quests.

larger groups of people. So a little research about an area you find yourself in can yield some inspiration. Missouri is known as the "Cave State," and the state of Tennessee has more than 10,000 caves. At Caverbob.com, you can find a list of the longest caves in the United States, and there are similar online resources for just about every country. Table 15 lists the ten longest US cave systems.

Table 15: Top 10 Longest Cave Systems in the United States

Name	State	Length (miles)	Protection Status
Mammoth Cave	Kentucky	412	National Park
Jewel Cave	South Dakota	209	National Monument
Wind Cave	South Dakota	154	National Park
Lechuguilla Cave	New Mexico	138	National Park
Fisher Ridge	Kentucky	130	National Park
Friars Hole	West Virginia	47	None
Binkley's Cave System	Indiana	44	None
Hellhole	West Virginia	42	None
Kazumura Cave	Hawaii	41	None
Double Bopper Cave	Arizona	41	National Park

Many of the longest caves in the United States are protected as either a national park or national monument, but several are not. One truism is that if there is one giant cave around, there are probably more little ones. If you are considering the ultimate in survival options for the next ice age, these might be good choices.

ABOUT LIMESTONE CAVES

Limestone is an excellent host rock for caves because limestone readily dissolves in water. In fact, the more acidic the water, the better for cavern building. Pushing slightly acidic groundwater repeatedly through limestone rocks invariably yields cave systems, sometimes miles in length. Such systems may be damp and contain running water, but they may also contain very large open areas with dry floors. Limestone cave systems may be ideal for temporary shelters or offer longer-term occupation if dry zones exist.

Minnetonka Cave in southeast Idaho contains fantastic shapes and characteristic limestone features.

Mines as a Shelter Hack

A slightly unusual hack for survival under extreme conditions is to consider old mines, which, after all, are essentially artificial caves. The problems with using mines as an emergency shelter are twofold. First and foremost, any mining district near a sizable town has long since been discovered by local teenagers and used as a party place. Broken glass, smashed cans, discarded personal items, old cars, used appliances, and even partially burned lawn furniture may greet your arrival. The ground is doubtless littered with sharp metal, nails, screws, and other tetanus shot–worthy junk. The urge to start picking up and bagging trash can be overwhelming.

Second, if partyers are too lazy to haul out their litter, they are likely to be too lazy to go out of the mine to relieve themselves, and the fumes can be overpowering. Enough said there.

As if those check marks against aren't bad enough, now consider the safety aspect. As the USDA Forest Service points out on its law enforcement page: "Abandoned mine sites are a great safety hazard. Many of these structures contain dilapidated frames, open shafts, and water-filled pits. The dangers that are found in the mines include old explosives, hazardous chemicals, bats, snakes, spiders, bobcats, mountain lions and other predators. Falls and cave-ins are common in these old mines."

Table 16 lists some common mine terms.

Table 16: Common Mine Terms

Term	Description
Adit	Tunnel that goes in horizontally
Shaft	Vertical pit, up to hundreds of feet deep
Stope	Carved-out opening inside the mine for removing ore
Pit	Open-air dig using bulldozers and dump trucks
Tailings	The waste rock dumped to the side, sometimes containing interesting minerals

Basically, there are two kinds of abandoned underground mines: those that required heavy timbering and support to maintain safety, and those that were carved out of solid rock. Generally, you don't want to explore or occupy the former. Wooden timbers can rot and become unstable. The sheer fact that they were necessary in the first place tells you that the miners didn't feel safe unless there was quite a bit of bracing. That probably indicates the mine has multiple fractures and faults running through it, and likely had numerous cave-ins even while miners worked it. That could even be the primary reason it was abandoned: Nobody wanted to work in it.

Stone pillars still support the roof at the Jubilee Mine near Nelson, Nevada.

The main problem is entropy, known to physicists as delta-S, or ΔS. It is sometimes defined as the degree of disorder or randomness in a system. In formal terms, entropy is the measure of a system's thermal energy per unit temperature that is unavailable for doing useful work. Because work is obtained from ordered molecular motion, the amount of entropy is also a measure of the molecular disorder, or randomness, of a system. German physicist Rudolf Clausius (1822–1888) described entropy using mathematics in 1865. He wrote that the energy of the universe is constant. The entropy of the universe tends to a maximum. In other words, the second law of thermodynamics states that entropy always increases over time, or, as the old poster warns, "Shit happens."

Applying that concept to being underground in a mine can be frightening. A sudden earthquake could unleash holy hell, causing ceilings to give way or floors to collapse. Mines constantly give off noises—good luck identifying them. Cornish miners exploiting the tin mines in their homeland believed in "wee little men [who were] . . . sent by the Romans to work as slaves," according to LegendsOfAmerica.com. Germans called them *Berggeister* and *Bergmännlein*, which translate to "mountain ghosts" and "little miners." European miners who immigrated to the United States in search of work in the Pennsylvania coal mines and riches in the California gold rush brought their stories and superstitions with them, and the legends quickly spread.

These mischievous imps weren't always portrayed as evil, according to the LegendsOfAmerica.com article. "The name 'knockers,' pronounced 'knackers,' comes from the knocking on the mine walls that often happens just before cave-ins. Actually caused by the creaking of earth and timbers, some thought these sounds of 'hammering' were malevolent, indicating certain death or injury, while others saw their 'knocking' as well-meaning, warning the miners that a life-threatening collapse was imminent. Yet others believed that the knocking sounds would lead them to a rich ore body and/or signs of good luck."

The worst mines to shelter in are coal mines, which are notorious for building up dangerous gases as the hydrocarbon-rich coal interacts with water. Mines that exploit ore associated with pyrite are nearly as bad, as water interacts with pyrite to create sulfuric acid, which can hang in the cold, misty air. The lowest areas are the worst, as the noxious air tends to be heavier and settles in the lower reaches. The Howden Group (howden.com), based in the United Kingdom, specializes in handling air and gas across multiple industries, and they characterize the worst mine vapors in table 17.

Table 17: Worst Toxic Gases Found in Mines

Gas	Formula	Characteristics
Methane	CH_4	Colorless, odorless, explosive
Carbon Monoxide	CO	Colorless, odorless, explosive, toxic
Carbon Dioxide	CO_2	Colorless, acrid smell, toxic
Hydrogen Sulfide	H_2S	Colorless, rotten-egg smell, explosive, toxic

Clearly, there are multiple hazards surrounding old mines, and they should only be an option in extreme situations. In addition to everything else, you could be trespassing. Removing any artifacts from these mines would be akin to stealing and punishable by heavy fines, even imprisonment. Most abandoned mines in the United States are gated or barred, preventing entry. Even that doesn't stop some determined explorers—but it should.

The US Geological Survey estimates there are about 65,000 abandoned hard rock mines in the United States. Since 1981, mine operators must post bond and rehabilitate their site when finished, but the vast bulk of properties were abandoned long before that law took effect.

The Upper Animas River drainage in Colorado is still stained red years after a disastrous iron-rich mine spill.

The worst of these properties have creeks running out of the mouth of the adit; this rust-colored water, laden with sulfides and heavy metals, poses great risk to wildlife.

As a survival hack, you definitely have to get lucky when considering an old mine for shelter. The best advice is to stay as near the entrance as possible. Use tarps, rocks, and trees to block the wind circulation, and move on the next morning.

The Mineshaft Gap

Near the end of director Stanley Kubrick's 1964 classic black-and-white film *Dr. Strangelove, or, How I Learned to Stop Worrying and Love the Bomb,* the assembled generals, scientists, and political leaders are coming to grips with the impending "Doomsday Scenario." A rogue US Air Force officer has unleashed a nuclear strike against the Soviet Union, which will in turn induce the newly installed (and never publicized) Soviet response of an automatic counterstrike, triggering total annihilation. The surface of the Earth is about to become a radioactive wasteland. The only possible response, according to Dr. Strangelove, the German-born scientist advising the president, is to hide out in underground mines for one hundred years, until the radiation fallout dissipates.

George C. Scott's character, General Buck Turgidson, soon perks up at the thought of repopulating the earth. He asks about the devastating consequences of abandoning the sanctity of monogamous relationships as Dr. Strangelove mentions that a ratio of ten women for every man should be required. The conversation then devolves into fears of a "Mineshaft Gap" if the Russians have already game-planned out this scenario and intend to mount a sneak attack to take over the US mineshaft space.

Multiple articles and reports of the time supported this fear. The *Bulletin of Atomic Scientists* in January 1959 included an article entitled "How Many Can Be Saved?" and discussed assumptions about the ability of the Russians to prepare civil defense plans. Political analyst Leon Gouré (1922–2007) prepared two key reports: *Civil Defense in the Soviet Union* (University of California Press, 1962) and *War Survival in Soviet Strategy: Soviet Civil Defense* (Center for Advanced International Studies, 1976). Both the United States and Soviet Union initiated programs that encouraged bomb shelter construction; the US National Archives has a reference to a short black-and-white film entitled *Walt Builds a Family Fallout Shelter,* complete with step-by-step instructions.

CONSIDER LAVA TUBES

Lava tubes are geologic features that form when hot, running lava forms a crust on top while the molten rock continues flowing down a gentle slope. Eventually the crust hardens and thickens, and if the liquid lava drains out and leaves a void, a lava tube results.

Such features come in all shapes and forms. Some are immense, cavern-like tunnels with a gradually sloping floor. Others are narrow, craggy, twisty, and tight—strewn with boulders and prone to cave-ins from above that allow light to enter. With the right water flow, sediment can build up at the bottom of the tube, drying to a wheelchair-compatible floor after the water stops.

Many volcanic terrains feature lava tubes. The composition of the lava is key; you need runny lava such as basalt that flows easily. Andesite tends to pile up—its namesake mountains, the Andes of South America, feature tall, steep crags. Rhyolite can be explosive, frothy, and prone to ash clouds.

Earth's lava tubes range in age from a few years old, such as those on the slopes of Hawaii's Kilauea volcano, to a few hundred thousand years or more. The freshest tubes have yet to be fully mapped, but in some areas they're known to stretch dozens of kilometers. In California's Lava Beds National Monument, for example, more than 800 lava tubes together extend at least 350 kilometers (approximately 217 miles).

Ape Cave, on the south side of Mount St. Helens, is a long, sinuous lava tube that is open to the public and easy to explore. The floor is wide and flat in many places, and the walls bear lines that mark repeated flows.

Kaumana Caves Park is a tropical lava tube near Hilo, Hawaii. An 1881 lava flow off the flanks of Mauna Loa created the 25-mile structure, but only 2 miles of it are open for public exploration. The rock is slippery and damp, and the cave itself is cold and dark. The floor is often obscured by boulders and is challenging to hike, but it's a nice relief from the hot humid air outside.

One of the more infamous lava cave systems was the basalt terrain of Northern California near Tule Lake. Now part of the Lava Beds National Monument, it was the site of "Captain Jack's Stronghold." This area witnessed a lengthy standoff between a local Native American tribe and the US Army in 1872–1873. A Modoc chief known as Captain Jack, whose given name was Kintpuash (or Kientpaush), meaning "Strikes the

This lava tube, known as Ape Cave, sits on the southern flanks of Mount St. Helens, with flow lines on the sides marking repeated pulses of runny basalt.

Kaumana Caves Park on the Big Island of Hawaii is cold and damp, at least near the front.

Water Brashly" held off up to 1,000 soldiers for several months. The warriors used the natural defenses as a fortress, resisting assaults by moving about the trenches and sheltering in shallow caves. During frustrating peace negotiations, Kintpuash shot General Edward R. S. Canby, who thus became the highest-ranking US soldier killed during the Indian Wars. The army soon captured and hanged the Modoc chief, but the monument remains as a testament to practical geology.

Planetary geologists speculate that the same events that happen on our home planet also occur on the Moon and on Mars, probably with an assist from a massive meteorite impact. Such extraterrestrial lava tubes would provide effective shelter against gamma radiation—a penetrating type of electromagnetic radiation that arises as atomic nuclei decay. Because it consists of the shortest wavelength of electromagnetic waves, gamma radiation packs the highest photon energy punch. Gamma waves are very dangerous, disrupting the cell structure of bone marrow and internal organs. Only the densest material, such as lead or thick layers of rock and concrete, is suitable protection against this deadly radiation.

Shielding protects against gamma rays but requires large amounts of material. Paper or even skin protects against alpha particles, which pack little punch, and aluminum shields against beta particles. Gamma rays are best absorbed by materials with high atomic numbers (*Z*) and high density, which contribute to the total stopping power. Because of this, a lead (high *Z*) lining is 20–30 percent better as a gamma shield than an equal mass of another, low-*Z* material, such as concrete, water, or soil. Lead has a higher density, but it is heavy and impractical to transport to another planet. Thus deep lava tubes with thick rock structures are potential spots for the living quarters and greenhouses that planetary homesteads will require.

Materials for shielding gamma rays are typically measured by the thickness required to reduce the intensity of the gamma rays by one-half (the half value layer, or HVL). Consider 1 cm (0.4 inch) thickness of lead to be the standard.

Table 18: Shielding Properties of Various Rocks Against Gamma Radiation

Material	Specific gravity (g/cm3)	HVL in centimeters	HVL in inches
Lead	11.3	1 cm	0.4 inch
Granite	2.4	4 cm	1.6 inches
Concrete	2.7	6 cm	2.4 inches
Packed soil, basalt	2.9	9 cm	3.6 inches

Table 18 is simplistic at best. This field is extremely complicated, and there are multiple equations for calculating exposure times based on shielding material—graphs that measure effectiveness for escalating intensity and the effect of combining shielding to combat alpha, beta, and gamma radiation. The key point is that solid rock is an effective shield, given enough thickness, and it is practical, given the expense of transporting material from Earth to the Moon or Mars.

Here on Earth, we have a built-in shield—the powerful magnetic field surrounding our planet, known as the magnetosphere. As explained at nasa.gov, the core of the Earth is also an electromagnet. Although the crust at the surface of the planet is solid, the solid core of the Earth is surrounded by a mixture of hot, molten iron and nickel. NASA explains that the result is a dynamo, a "machine" that converts mechanical energy into electrical energy like a generator: "The magnetic field of Earth is caused by currents of electricity that flow in the molten core. These currents are hundreds of miles wide and flow at thousands of miles per hour as the earth rotates." The result is a powerful magnetic field that passes through the crust and into space, forming a shield.

Earth's moon has no measurable magnetosphere today. Apollo astronauts that landed on the Moon were outside the normal protection of Earth's magnetic field and reported headaches, cataracts, and flashes of light, and were assumed to be at a higher risk for cancer. Occupants of the International Space Station require extensive shielding for protection against radiation; space walks in relatively thin suits are extremely hazardous for long periods. Even more deadly would be a sudden solar flare, which gives off tremendous pulses of radiation across all spectrums.

The Environmental Protection Agency has a good write-up for the dangers of solar flares at epa.gov: "Solar flares are large eruptions of energy coming off the Sun containing several different forms of energy: heat, magnetic energy, and ionizing radiation. The ionizing radiation released during solar flares includes X-rays and gamma rays. These rays of ionizing radiation can damage satellites because they are in space and are not protected by the Earth's atmosphere. Magnetic energy from solar flares can interrupt radio communication on Earth or damage communications satellites."

Thus solar flares are the planetary equivalent of one of the most dreaded (and likely) doomsday scenarios: electromagnetic pulse (EMP) bombs. An EMP attack would theoretically bomb us back to the Stone

Age; all electronics would be at risk of frying, and our computer-driven lifestyle would be at severe risk. Telecommunications systems would go down, electronic commerce would be disrupted, and marauding bands of pillagers might not be far behind.

Unfortunately for future Mars homesteaders, that planet has a much smaller magnetosphere than Earth. According to Rafi Letzter, posting for the May 11, 2020, edition of *LiveScienceDaily*, Mars colonies may have to rely on lava tubes for lengthy stays. "A team of researchers has identified what could be future Martian explorers' best possible hide-out: a string of lava tubes in the low-lying *Hellas Planitia*—an impact basin blasted into the Red Planet's surface by ancient meteor impacts," Letzter reports. He cites the work of Antonio J. Paris and a team of scientists at the Center for Planetary Science (planetary-science.org).

The Mars Global Cave Candidate Catalog (MGC) at astrogeology.usgs.gov provides latitude and longitude coordinates, feature type, priority (confidence) rating, and a brief comment about each candidate. Types of cave entrances identified in the catalog include lava tube skylights, deep fractures, atypical pit craters (APCs), and other details. Most entries in this catalog were identified through surveys of images from the Mars Reconnaissance Orbiter's (MRO) Context Camera (CTX) and High Resolution Imaging Science Experiment (HiRISE) cameras.

CHAPTER 5

STAYING WARM

If you're familiar with the concept of Maslow's Hierarchy from that long-ago freshman class in psychology, you might recall there is a consistent pattern to the steps we take from survival to self-actualization. In his 1943 paper Abraham Maslow (1908-1970) placed at the beginning, or bottom of his pyramid the two key levels for survival—first, the physiological needs for warmth, rest, water, and food. The next level was security and safety, but we're not done with the basic needs yet.

THE IMPORTANCE OF FIRE

Fire has long been a vital resource for humans struggling to survive. In addition to providing heat, fire breaks down proteins and carbohydrates, making them easier to consume and digest. It also serves as deterrence to predators hoping to score an easy meal. Fires can clear off brushland to make the land more amenable to agriculture, it can harden sharpened sticks into better weapons, and it can convert clay into hardened bricks and pottery.

There is evidence of widespread fires in the fossil record. According to a 2018 article by Andrew C. Scott at Time.com, the oldest evidence of fire comes from charcoal found in rocks formed about 420 million years ago during the late Silurian period. "The first extensive wildfires recorded came somewhat later, dating from around 345 million years ago, the early Carboniferous period," Scott added. So we can theorize that fire has been a constant feature on Earth, just waiting to be harnessed.

In his epic book *Sapiens* (HarperCollins Publishers, 2015), Yuval Noah Harari writes that our species took quite some time to reach the top of the food chain. "For more than 2 million years, human neural networks kept growing and growing, but apart from some flint knives and pointed sticks, humans had precious little to show for it," he argued (Harari, 9).

Then things changed quickly. "A significant step on the way to the top was the domestication of fire. Some human species may have made occasional use of fire as early as 800,000 years ago," Harari writes. "By about 300,000 years ago, *Homo erectus*, Neanderthals, and the forefathers of *Homo sapiens* were using fire on a daily basis. Humans now had a dependable source of light and warmth, and a deadly weapon against prowling lions. Not long afterwards, humans may even have started deliberately to torch their neighborhoods. A carefully managed fire could turn impassive barren thickets into prime grasslands teeming with game. In addition, once the fire died down, Stone Age entrepreneurs could walk through the smoking remains and harvest charcoaled animals, nuts and tubers" (Harari, 12).

A December 2021 article at NewScientist.com describes work done by Katharine MacDonald at Leiden University in the Netherlands that reinforces Harari's point. Her team found evidence at Neumark-Nord in Germany that Neanderthals were also able to harness fire to change their environment. The ancient humans apparently cleared off forest landscape around adjoining lakes to improve hunting opportunities.

Harari pieced together some interesting evolutionary traits thanks to fire. He wrote that while chimpanzees spend up to 5 hours per day chewing raw food, humans get the same job done in 1 hour. Staples that are indigestible raw, such as wheat, rice, corn, and potatoes, now could stock the larder and join the dinner table. The result on our bodies soon showed up in the fossil record in the form of changes to our teeth and additional internal changes that remain with us today. "The advent of cooking enabled humans to eat more kinds of food, to devote less time to eating, and to make do with smaller teeth and shorter intestines," Harari noted. He added that many scholars believe that since the brain and the digestive tract are both huge energy consumers, once the intestines shrank, the brain could expand and the use of tools could explode.

We can guess that the first exploitation of fire happened by accident, probably "recycled" from a lightning strike that smoldered for hours or days. Early humans apparently learned fairly quickly how to coax glowing coals into flame, how to collect the right tinder, and how to transport embers using dried fungi. Without knowing the exact science, they learned that there are three crucial ingredients to a fire: fuel, oxygen, and a spark.

THE HUMBLE FIRE RING

Building a proper fire for warmth relies on both preparation and science, plus a little geology. Laying out sticks and tinder in a haphazard fashion on cold, damp ground without a solid base of resources at the ready often leads to disaster. Consult any of the standard references if you need help with the basics of tinder, fuel, and flame—we'll instead concentrate on the geology behind safely harnessing fire, starting with a ring of rocks.

Modern campers in developed campgrounds benefit from the latest in fire ring technology, with an impressive set of features such as movable grates, air holes, tall sides, and either rectangular or circular construction. We'll talk more about fireplaces, hearths, and kilns later; for now, a simple circle of stone is a key resource dating back to our early roots.

Tidy fire ring with nearby boulders for seating plus bare earth except for the remains of the wood pile, which was probably too close

Modest fire ring with massive boulder for a windbreak and a decent view

In a true survival situation, the urgency to get warm is a driving force, especially in extreme cold. Just know that you don't always need a fire. The best advice to follow in most camping situations is "Leave No Trace." Building, tending, and extinguishing a fire consumes a lot of time and energy—and definitely leaves behind evidence. That may be a good thing if you are hoping to be tracked and located, but many times, fires can be more hassle than they are worth.

With wildfires plaguing the western states in particular, and knowing that humans cause most of the problems by far, it's worth weighing the advantages and disadvantages of making a fire. Add to that the fact that you can be on the hook for fines and damages if your fire is proven to be the cause of property destruction or loss of life, and we should probably all think twice about firing one up. In 2020 a study by researchers at the University of Colorado at Boulder cited human causes for 97 percent of all home-threatening wildfires. That category includes arson, debris fires, equipment malfunctions, and more, so there's no doubt that "campfire season" has taken on a whole new meaning.

Still, if you do find yourself in need of a campfire, either in an emergency or when conditions and regulations allow, there are some easy geology concepts to master. The first involves using rocks as a fire ring, a time-honored human tradition.

There are several good reasons for encircling a fire with rocks. First, it keeps the coals from wandering far from the flames. Place your ring well free of overhanging branches. Inexperienced campers often fail to take into account the zone above the flames. In Jack London's classic short story "To Build A Fire," a greenhorn blunders out in -50°F weather during the Alaska gold rush and falls through the ice while crossing water. His emergency fire under a tree blazes brightly for a minute before the branches laden with snow above him thaw out and release their burden. His dog survives; he does not.

Fire marshals like to see a bare-earth safety zone of at least 10–12 feet in diameter around your stone circle. That prevents most of the issues caused by flying sparks landing on combustible dry leaves, dead grass, dropped needles and cones, etc. Get the fire started first, then, while tending to the flames through the initial period of catching and settling into consistent flame, set about clearing the rest of the zone. It's easy enough to scrape clean a safety zone; in some jurisdictions you need one to avoid a fine, along with an approved fire ring, a bucket of water at the ready, and a stout shovel standing by.

No special tools are required to clear the top layer of soil, leaves, and rocks—you can use the side of your boot or shoe and slide it in short strokes like a makeshift rake. That motion can be hard on moccasins or running shoes, but boots do a good job. Pile up the debris as a ring and continue to push it away from the center. This provides your camp with a more civilized look and is a big safety improvement, as you're less prone to roll an ankle on some stray, perfectly round pebble.

If campers in your party insist on going barefoot, you're less likely to be applying first aid on tender feet with a proper safety zone. A handy additional safety precaution is to use a very strong magnet around existing fire rings. You'll be surprised at the number of rusty nails, bottle caps, screws, and other bits of metal you'll pull from an old ring, especially at a developed camp. If nothing else, you can show off the bounty to young children and encourage them to put their shoes back on or risk the need for an unpleasant tetanus shot.

The amount of metal trash around established campfire rings can be a tetanus shot waiting to happen.

Note also that some localities now have regulations about the size of your fire. British Columbia, for example, expressly forbids bonfires; thus fires there should be no higher and no wider than 0.5 meter (1.6 feet).

In many emergency situations, you may be faced with extreme wind conditions, and there's a geology hack for that too. Stacking up suitable rocks creates a windbreak, which can help conserve your fuel supply. You can pile up clay-rich mud or construct wooden structures too. Strong winds will devour your wood pile rapidly by supplying so much fresh oxygen that the fuel is greedily consumed. Creating a low stone wall, chinked with soil, can remedy that. Flat rocks such as platy andesite, schist, and slate make excellent building supplies for battling the wind.

The rocks in a fire ring allow you to get good angles when you place fresh wood on the fire. When using the log-cabin style for setting the wood, stacking them in a square pattern, or a tepee, with the sticks standing against one another, the rocks help with construction. If using the star pattern, with crisscrossed firewood, the rocks can help keep the ends tilted. If stuck in a situation where your wood supply consists of very long poles or branches and you don't have the tools to break them down into suitable fuel, you can brace them with rocks to keep them burning through the middle. You can also use rocks to break the logs.

Additionally, rocks in a fire ring help store and radiate heat. One method for constructing a fire ring emphasizes digging out the pit about 6 inches, then lining the sides with rocks to heat them. In extremely cold weather, you can carefully remove and bury those rocks under plenty of soil where you sleep. The caution there is obvious: If the rocks are too hot, they can start a fire—or melt nylon.

Exploding Rocks

This is not a myth—rocks do occasionally explode when heated quickly, as trapped water quickly turns to steam with no escape path. When constructing a simple fire ring, you can practice some simple geology before you start. In general, hard rocks like granite, gneiss, marble, or slate are very dense, and therefore less likely to absorb water and explode when exposed to heat. Other materials that are safe to use around and in your fire pit include fire-rated brick, lava glass, lava rocks, and poured concrete. The most prone to explosion seem to be basalt, shale, and any rock that's been sitting in or near water.

At DecideOutside.com, there's a good explanation about the problem of exploding rocks, which the author attributes to two main issues:

1. Waterlogged rocks that get hot quickly build up steam pressure inside, which releases in an explosion. The rocks most prone to this safety hazard are damp, round river cobbles.
2. Rocks with different densities, such as a porous zone and a tighter zone. If the rock heats and expands at different rates, it can shatter.

Generally, if your fire builds slowly and doesn't get hot in a big hurry, you should be safe. The gradual buildup of heat allows steam to escape at a leisurely pace. If you progress rapidly from cold and damp to fiery red-hot rocks, you might have issues. If the rocks in your fire ring start exploding, dial back the fuel. Most of the time the rocks will just burst, without much drama. But sometimes the flying chunks of rock come ripping out of the hot flames at very high speed. Your best bet is to try to cool off the fire and use a shovel or large stick to move the fire ring rocks back another foot.

ANCIENT FIRE STARTERS

In most scenarios, you should be able to pile up a few rocks, gather some firewood, grab some easy tinder from the dry zones near a tree trunk, and prepare for ignition with matches or a lighter. But how did ancient humans get a spark? In the first part of her classic "Earth Children" series that began with *The Clan of the Cave Bear*, author Jean Auel set her main character Ayla down in a toolmaking session, absentmindedly flaking chert into a tool. Ayla got her hands on a brassy, shiny rock and whacked it against the chert, resulting in a shower of sparks.

Some of these pyrite specimens are probably too good as specimens to ever crush against a striker.

Thus, in Auel's telling, was born the firestone that freed early humans immensely, and her vignette sounds logical. Stone Age rockhounds must have spent considerable time seeking interesting material, experimenting endlessly, and searching and exchanging resources. And pyrite is the most common sulfide mineral.

An article at Research.Reading.ac.uk titled "Finding the fires of early humans," written by Rebecca Scott, Marc Curtis, and Robert Hosfield, describes evidence for the use of fire in the archaeological record. They guess that pyrite nodules were a crucial component of fire-starter kits, at least in Europe and Greenland. Of course, wooden fire starters such as bow drills would not be easily preserved and may have been an equally important early invention. But were pyrite nodules readily available to prehistoric practical geologists?

In the January 2, 2015, edition of *Antiquity*, Dick Stapert and Lykke Johansen discuss finding flint implements with rounded ends excavated at several Upper Paleolithic sites in Denmark and Holland, which they interpreted as "strike-a-lights used in combination with pyrites." Authors Alastair H. Ruffell and David J. Batten describe the Cretaceous dry period that resulted in "abundances of carbonate-rich sediments, red beds, firmgrounds, carbonate nodules and pyrite concretions." A 1989 paper by Peter J. MacLean and Michael E. Fleet published in *Economic Geology* provides evidence for stream deposits of pyrite in South Africa, near the Witwatersrand gold fields, and there are fantastic, perfectly formed pyrite cubes still recovered in Cretaceous marl beds rich in calcite at Navajún, Spain.

So clearly, pyrite is geologically abundant enough for ancient rockhounds to have found it, and we have ample evidence that it was exploited. The science behind starting a fire by banging the right two rocks together is actually based on the Mohs hardness scale. Chert, a common form of quartz that typically develops in seams and pods within limestone, has a hardness of 7 out of 10; it will scratch just about anything except extremely hard minerals such as jadeite, topaz, corundum, and diamond.

When you violently strike a hard quartz material such as chert against pyrite at the correct angle, you shave off small particles of the pyrite. Because the pyrite particles are superheated to a molten state due to the pressure used to break them off, they oxidize and create sparks.

MODERN FIRE STARTERS

Modern fire-starter kits usually substitute steel for pyrite, but you need the right kind of steel. Typical low-grade steel has a hardness of 4–4.5, while specialty hardened steel can be harder than chert, reaching 7.5 or even 8 on the Mohs scale. At that hardness, you'd remove fragments from your chert, not from the metal. A high-carbon tool-grade steel, water-quenched, should suffice, as it won't reach 7 on the Mohs scale. Most decent metal files will also work, but higher-end files may also be harder than your chert.

A classic, portable piece of steel called a fire starter dates back to the beginning of the Iron Age in design. It is curved to (somewhat) protect the knuckles, and once tiny bits of steel spall off after striking, they are capable of igniting fire consistently over a lifetime of use. Because of their high level of dependability, these strikers are popular with outdoorsmen, survivalists, preppers, and those practicing self-sufficiency. They are the perfect size for portability, measuring only 2.0 × 3.7 inches, so are ideal as an emergency fire starter for backcountry wilderness survival. You can strap it to your bag for added peace of mind, knowing you'll always be able to create sparks with flint, chert, quartz, jasper, agate, or other easy-to-find rocks.

This handy iron striker fits in your hand across the knuckles, making it easy to whack on chert, flint, etc., for a spark.

Keep in mind that it takes practice to learn what materials work best, how to strike, and how to avoid hurting yourself when you begin to tire and err with a powerful attempt. Practice a few times before you need to rely on this ancient method for survival.

Carbon steel strikers create sparks by shedding molten microparticles of steel after striking, but to ignite a fire you'll need something to catch the sparks, such as char cloth or fluffy tinder, then patiently and cautiously blow it into a flame.

Fire starters today are one of two types: ferro rods or magnesium rods. At CrisisEquipped.com, there's an excellent discussion on the pros and cons of the two different tools. They start with the ferrocerium ("ferro") rod first, also known as fire steel. By and large, all ferro rods contain a mixture of metal alloys, including cerium, iron, magnesium, and various rare earth elements (REEs) such as lanthanum, praseodymium, and neodymium. The manufacturer coats the rods with a black coating, which you must scrape away to reveal the silvery metal inside.

Since shaving off the magnesium into a pile is an important step, you might think of purchasing pre-shredded shavings and storing them in an airtight plastic bottle. Then all you need is the ferro rod for sparks.

Be sure to spend time making a plan before you spark up, with tinder, small sticks, and larger sticks all ready to go. Char cloth, created by burning cotton fabrics in a metal container with a very small hole to control airflow, is a great tool for catching the spark from a ferro rod, but it doesn't burn hot like magnesium. Dry tinder from an empty bird's nest, dry bark or moss, dryer lint, or similar material is essential to feed into your flame.

CHAPTER 6

SAFE WATER

At this point, you've come a long way in your hypothetical survival situation—you're making good decisions, you can breathe, and you are temporarily safe from the elements so that maintaining your core body temperature is no longer an immediate issue. For now, your next most critical need is likely to be water. If you have access to a decent water source and a pot, you can boil the water to kill off microorganisms and safely move on to acquiring food. But it's never that simple, is it?

DEHYDRATION IS DEADLY

First, a few words about the dangers of running out of clean drinking water. The challenge is to avoid dehydration and its deadly cousin, heatstroke. You should know and understand the symptoms for dehydration, as explained by the United Kingdom's National Health Service:

- Feeling thirsty
- Seeing dark yellow and strong-smelling urine
- Feeling dizzy or lightheaded
- Feeling tired
- Noting a dry mouth, lips, and eyes
- Urinating little, and fewer than four times a day

Several factors can increase your odds of dehydration:

- Diabetes
- Vomiting or diarrhea
- Sun exposure
- Alcohol intake
- Sweating too much, especially after exercising
- Body temperature of 100°F (38°C) or more
- Medicines such as diuretics

The treatment for mild dehydration is simple: Get out of the sun, cool off, and drink fluids. Intravenous drips to replace fluids are crucial for serious cases. Electrolytes help replace lost minerals, so energy and sports drinks work well, as do products like Pedialyte for children. Failure to treat the symptoms can lead to an even bigger problem: heatstroke.

Dehydration can quickly lead to heatstroke, with the following symptoms:

- Heavy sweating
- Cold, pale, and clammy skin
- Fast, weak pulse
- Nausea or vomiting
- Muscle cramps
- Tiredness or weakness
- Dizziness
- Headache

Resolving heatstroke usually requires intervention from trained medical personnel. Sufferers require immersion in cold-water baths, intravenous fluid replacement, and other drastic measures.

AVOID EATING SNOW

We all know the adage to avoid yellow snow. But should you eat any snow, period? The answer is, like most scenarios, "It depends."

As ice melts, the molecules change their state from a solid to a liquid. Every state change is accompanied by either a release or accumulation of energy, and in the case of melting from a solid to a liquid, a total of 334 joules of energy are required to melt 1 gram (0.035 ounce) of ice at 0°C (32°F), which is called the "latent heat of melting." That energy comes directly from your body's core temperature if you melt the snow in your mouth, and over time, those joules can add up. That goes for placing a plastic bag of snow in your armpit to melt too. Instead, get a fire going and melt the snow in a pot like Arctic and Antarctic explorers.

Of course there are no absolutes. In 2018 a California grandmother was stranded in her car for ten days after taking a wrong turn during a mountain storm. She survived by eating snow and wrapping herself in extra clothes to keep warm. Interestingly, she violated another common rule in survival: She left her vehicle. Most experts advise remaining in the car, as it is easier for rescuers to find. This hardy grandmother set out on her own, hiking about a mile to self-rescue.

Snowflakes may look innocent, but there are dangers to consider before eating snow.

Besides the energy required to convert ice to water, there is another good reason to avoid snow as a water source: It generally carries pollutants. At Sciencing.com in 2017, Carolyn Csanyi wrote a very clear explanation about the beginning moments of snowflake formation: "Water droplets in the cooling air mass condense around tiny particles in the atmosphere, such as soot, pollen, dust or dirt."

John Pomeroy, a researcher who studies water resources and climate change at the University of Saskatchewan, was quoted in a 2015 article at NPR.org saying it's better to wait until a few hours into the snowfall to gather your fresh catch. Snow acts like a kind of atmospheric "scrubbing brush," he explained. After the snow has been falling for a while, it will have scooped out most of the atmospheric pollutants.

In the same web article, Staci Simonich, a professor of environmental and toxic ecology at Oregon State University, reported that she found pesticides that were thirty, forty, and fifty years old at high elevations in several US national parks. She tested snow in the Olympics in Washington, at Mount Denali in Alaska, and in Sequoia National Park in California. Although even those levels were one hundred times lower than what's deemed safe for drinking water, it's still a risk.

Fog Harvesting

Some of the harshest deserts in the world manage to host a wide and diverse ecosystem because animals take advantage of morning dew and fog. If you have an extremely absorbent towel in your kit, such as the legendary "ShamWow" towel, you can wipe the dewy leaves in the morning and squeeze out the water into a container. Note that you might have to treat the water by boiling or disinfecting, as there could be animal waste on the leaves.

A loma ("fog oasis") is a temporary oasis born from mists in the Peruvian desert at the edge of the Pacific coast. In the Preceramic period, 8000–1850 BCE, vegetation flourished in these microsystems, and today they are rich archaeological resources. Most of the plants in a loma store excess water in roots, corms, tubers, and bulbs, according to David Beresford-Jones and his team in an article published in *Quaternary Science Reviews* in 2015. Such bulbs and tubers would be another source of water for surviving in the desert.

In Morocco, a "fog farm" built in collaboration between Aqualonis and other German companies covers three football fields and collects more than 37,000 liters of water on a good day using "CloudFisher" billboards. The collectors use a fine-meshed black polymer stretched between steel poles, with a drip-catcher at the bottom of the mesh to collect the water.

You might be able to turn the fine mesh of your tent windows into a similar device. Or you might be able to locate rocks doing the same work. According to an article for Columbia Climate School at news.climate.columbia.edu, the technology is ancient. "Fog or dew collection is an ancient practice. Archaeologists have found evidence in Israel of low circular walls that were built around plants and vines to collect moisture from condensation. In South America's Atacama Desert and in Egypt, piles of stones were arranged so that condensation could trickle down the inside walls, where it was collected and then stored."

DRINK YOUR OWN URINE?

At FunnyorDie.com, there are eleven hacks for surviving a nuclear war—one of the most feared apocalyptic scenarios for survivalists and preppers. The hacks, generally not serious, include increasing your lead intake to guard against radioactivity, gentrifying a nearby planet after fleeing our own, and this nugget:

Drink small amounts of piss so when it's all you can drink it's not so bad.

Many claim that in a survivalist situation, drinking your own urine can stave off dehydration. In fact, this advice is risky. Not only will your urine barely rehydrate you, it could have the opposite effect and dehydrate you at a faster rate. Many survival experts believe people who drink their own urine and survive are simply lucky.

Of course there are exceptions. A Chinese man trapped under a piece of ceiling managed to survive for more than six days after the massive earthquake in Sichuan province in 2008—in part, reports the *Wall Street Journal*, by drinking his own urine, which he collected using leaves and rubbish.

Here's the science, however. What you are accomplishing when you drink your own urine in a survival situation is giving yourself more time, perhaps an extra day or two, so it's a completely short-term solution. Typical urine is about 95 percent water and sterile, so in the short term it's safe to drink and does replenish lost water. Seawater, by contrast, is about 96.5 percent water. So drinking your urine is just a little less safe than drinking seawater.

In a 2015 article at Quartz India (qz.com), Mridula Chari cited several Indian politicians and dignitaries that extolled the virtues of consuming human urine. While debunking the oft-repeated myth that Mahatma Gandhi drank his own urine (he did, however, advocate consuming cow urine), she noted that an ancient Sanskrit text, Damar Tantra, exalts the practice.

The problem is that the other 5 percent of urine is waste products, including nitrogen, potassium, and calcium. And in a dehydrated human, that 5 percent can range toward double digits. So now all the chemicals your kidneys worked so hard to remove are right back into your digestive system. The more salt chemicals you ingest, the more water will be drawn out of your cells through the process of osmosis. After too many cycles, your ever-darkening urine will likely bring on kidney failure.

So at best, drinking your urine is a "break glass in case of emergency" strategy. Aron Ralston, the man who amputated his own arm with a pocketknife to escape from a boulder that pinned him in a Utah canyon in 2003, said he was forced to drink his own urine. He didn't like it, but we know that wasn't even the most drastic step he took. The Army Field Manual, available as a PDF if you search for it, notes that in the case of crushing injuries that damage muscle fibers, your body will secrete even more potassium and phosphorous into the bloodstream, which the kidneys will struggle to filter out.

DISTILLING WATER

What you can do with urine is distill it. Most survival books show a simple solar still; the process is fairly simple. There are multiple hacks for harnessing evaporation to recover drinkable water. The most common one in older survival books is the solar still:

1. Dig a shallow pit, level on the bottom, about 1 foot in depth, lined with clay if possible.
2. Pour in liquids—seawater, urine, crushed plants, etc.
3. Place a small cup or saucer in the middle, elevated above the "soup."
4. Cover with a clear plastic sheet.
5. Place a small stone in the middle of the sheet directly over the cup.
6. Cover the edges with extra soil or sand to seal it completely.

This solar still should heat up quickly, evaporating off pure water. The water will condense on the sheet and slide down to the middle, to drip into the cup. The process is very slow—a couple tablespoons of pure water per day is a good yield. Since plastic bag waste is common around ocean islands, this is a practical solution for tropical survival.

There's also a hack using another ubiquitous form of garbage—the plastic bottle method. There are several variations, but here's one:

1. Remove the labels and fill one large plastic bottle with dirty water, seawater, etc., to about one-quarter full.
2. Join a second plastic bottle to the first one, mouth to mouth, using sturdy tape. You will have to be able to repeatedly break and reseal the connection.
3. Set the bottles in the sun, with the empty bottle elevated a couple inches.
4. Allow to heat in the sun for a few hours at least. The pure distilled water vapor rises and the gas fills the system. As the water condenses it mostly pools in the shoulder of the upper bottle. Ideally the bottle should be UV-stable and free of bisphenol (BP).

A third hack is a little more complex:

1. Cut off the bottom of a large plastic container, such as a 1-liter water bottle. Leave the lid on.
2. Fold the rough edges of the cut back into the plastic to form a collar or gutter about 1 inch in size. Take care to not split the plastic

when folding it in, as the gutter will be catching the distilled water.

3. Cut off the top of a standard 11-ounce beer or soda can.
4. Fill the can with the source water.
5. Place the can with the source water on the ground, as level as possible.
6. Place the larger plastic bottle snugly over the source can.
7. Allow the system to sit in direct sunlight for several hours.
8. Gently remove the upper bottle, taking care to save the pure water in the collar reservoir. You can tilt the water bottle so the pure water is trapped in the mouth next to the cap, then pour into a clean container.

You would need at least a dozen of these contraptions to get you through a day—maybe two or three dozen. The advantage is that they are made of readily found materials—garbage that washes up on just about every ocean beach. Cutting the plastic bottle and the aluminum can will require a knife or a sharp rock; we'll talk about fashioning your own primitive tools later.

Humans have always faced challenges over water. The earliest camps and communities typically situated near water sources, usually springs and creeks in the beginning, then streams and rivers as water needs increased. Our ancestors quickly understood the basics of taking water upstream and flushing wastewater downstream. That worked great until there was another group downstream.

Researchers can't know for certain, but it's likely that early humans soon made the connection between bad water and pathogens that cause disease. It's also likely that their stomachs were more adapted to consuming marginal water, similar to the way range animals are able to drink freely from suspect water holes.

The likely earliest known permanent settlement was at Jericho, in the Jordan Valley to the east of Jerusalem. Now a protected site, archaeologists guess that Jericho was founded about 10,000 years ago, or about 8000 BCE, when early hunter-gatherer clans began staying longer at the nearby Ein es-Sultan spring. Scientists have determined that the spring was a popular camping ground based on scattered remains of crescent-shaped Stone Age tools. Around 7600 BCE a long period of drought ended, and full-time habitation began. Fortified walls described in the Old Testament date to the Pre-Pottery Neolithic period, around 6000 BCE.

At Haaretz.com, researchers describe evidence of a well dug 6,500 years ago in the Jezreel Valley of Israel. Yotam Tepper, the Israel Antiquities Authority archaeologist who supervised the excavation, pointed to shards of flint as evidence of the simple stones used as tools. "The well shows the . . . ancient inhabitants of the area [used] know-how of the geology and hydrology of their environment" to dig deep enough to reach the water table.

In Egypt there are traces of wells, and in Mesopotamia of stone rainwater channels, from 3000 BCE. From the early Bronze Age city of Mohenjo-Daro, located in modern Pakistan, archaeologists have found hundreds of ancient wells, water pipes, and toilets. The first evidence of the purposeful construction of bathrooms, toilets, water pipes, and drainage systems in Europe comes from Bronze Age Minoan (and Mycenaean) Crete in the second millennium BCE.

Domestic wastewater disposal has involved reuse in fields in the Middle East, India, and Greece since the Bronze Age, from 3200 to 1100 BCE. In his book *Sanitation, Latrines, and Intestinal Parasites in Past Populations* (Routledge, 2016), Piers D. Mitchell provides a brief, scholarly history of sanitation. He notes that early hunter-gatherers lived in very small groups and moved around frequently, facing little risk from parasites and illness as long as they kept moving and cooked their food thoroughly.

Once cities began to grow, individual households began investing in their own sanitation solutions, using deep-pit and sloped-drain toilets. "By the 3rd millennium BCE baked-brick toilet seats started to appear, and by the 2nd millennium BCE the design started to allow flushing with water," Mitchell writes. By 2000 BCE, ancient Sumerians had developed clay sewer pipes. By the fourth century BCE there were toilets across Greece with anatomically shaped seats in both public and private buildings, Mitchell continues.

Early Roman water systems frequently used lead pipes, as people were unaware of the dangers of lead poisoning. Lead gets its symbol Pb from the Latin word *plumbum*, and the term "plumber" stuck for people who work with all manner of water pipes.

This lead pipe excavated at Pompeii shows just how widespread the ingestion of lead was in ancient Rome.

ALL ABOUT FLOCCULANTS

If you seek to clear large amounts of water in a settling pond or cistern, there is a geology hack for that. It involves the use of a flocculent.

Flocculants, or flocculating agents (also known as flocking agents), are chemicals that promote a process by which colloidal particles come out of suspension to become sediments.

Colloids are a noncrystalline substance consisting of large molecules or particles. Colloids include gels, sols, and emulsions; the particles do not readily settle, and they cannot be separated out by ordinary filtering or centrifuging like those in a suspension. In terms of water purification, think fine clay particles.

Flocculation causes suspended particles in liquids to clump together, forming a floc. The science is ingenious: The unwanted particles have a negative charge, so the positively charged flocking agent, such as alum, neutralizes them during coagulation. Then the particles are drawn together by the van der Waals force, forming floc.

A tablespoon or half an ounce of alum will purify up to 20 gallons of water. If you have a small, cloudy cistern that holds about 100 gallons, you can do the math: 5 × 0.5 ounce = 2.5 ounces. Consider the challenge of clearing the water in a small pond on your bustling, growing homestead. Experts recommend 25 to 50 pounds of alum per acre-foot of water, which should clear the water body in a few hours. If unsuccessful, try another 25 pounds.

Alum crystals are fairly easy to grow if you want to give the kids a fun lab experiment.

So, how do you calculate the amount of cloudy water to clear using alum? An acre foot of water equals about 326,000 gallons, or enough water to cover 1 acre of land 1 foot deep. To put it another way, 1 acre foot of water is enough to flood most of a football field to a depth of 1 foot.

An acre was historically defined as 1 furlong by 1 chain, or 660 feet by 66 feet. Thus, the square footage is 43,560 feet. That's about 4,840 square yards, so now we can just do the math.

Calculating Gallons in a Pond

Say you have a really dirty pond roughly 50 feet long by 20 feet wide. Other than the sediment, it should be clean water (in this hypothetical example the livestock area drains to a different topological unit). That's about 1,000 square feet of pond area. You can measure the depth at several places and take a rough average; say it's 10 feet deep at the deepest part, 1 foot deep at the edge, but generally about 5 feet deep, with not much shallow area. You might say it averages about 6 feet deep.

1. Multiply that average by 1,000 square feet, and you get 6,000 cubic feet.
2. Convert cubic feet to gallons by multiplying the cubic feet by 7.48. Your new total is 44,880 gallons, which is about ⅐ acre foot—44,800/326,000 = 0.1374, or about ⅐.
3. Multiply ⅐ against 25 pounds of alum per 1 acre-foot to get about 3.5 pounds of alum for your pond. If that doesn't work after a few hours, try another dose.

This isn't an exact science, and since alum can cost as little as $4.99/pound online, you have a fast and fairly inexpensive way to clear the water of suspended particles.

Can You Make Your Own Alum?

Transforming aluminum-rich shales into alum crystals is a complicated process. In his book *The Floating Egg: Episodes in the Making of Geology* (Jonathan Cape Books, 1998), Roger Osborne describes the challenges and methodology used to make alum.

The first stage in this process was to quarry an aluminum-rich shale out of the ground. To produce 1 ton of finished alum you would need up to 100 tons of shale.

The next stage was to burn the shale in large piles called clamps. These clamps could be up to 100 meters long and 30 meters tall and burned for as long as twelve months. The burnt shale was then steeped in freshwater to draw out all the useful chemicals. The liquid then went to the alum house—this is where the transformation happened.

After evaporating off the water, chemists added concentrated human urine shipped in from major cities, although burnt seaweed ashes also worked. This liquid was then cooled and allowed to crystallize slowly.

At this point it was crucial not to let the liquid cool too much, as ferrous sulfate would also crystallize out and pollute the alum. To prevent this, the alum makers added a fresh hen's egg to the solution. Once all the alum had crystallized, the density of the liquid changed and the egg floated to the surface. At this point, all the remaining liquid was drained off and the pure alum could be washed and packed.

HOW TO DISINFECT WATER

The EPA recommends using regular, unscented chlorine bleach products for disinfection and sanitization, as indicated on the label. The label may say that the active ingredient contains 6 percent or 8.25 percent of sodium hypochlorite.

Sodium hypochlorite is NaClO. Think of it as a sodium salt of hypochlorous acid. (It's a relatively cheap product, but there are hacks to get around bleach via substitution, which you can find at Survivorpedia under "DIY Bleach at Home.")

The first method is to dilute the product used to blast a swimming pool or hot tub into submission: "pool shock." The powder ships at around 70 percent calcium hypochlorite. Mix 2 tablespoons of pool shock with 3 cups of water and let sit until the solution is no longer cloudy. This solution is now equivalent to 6 percent bleach in the calculations below.

Or you can mix up some peracetic acid (also known as peroxyacetic acid, or PAA), which has the formula $C_2H_4O_3$. According to the EPA, quoted at Survivorpedia, peracetic acid kills *E. coli*, salmonella, listeria, staphylococcus, shigella, and other bacteria.

To mix it up, you need to combine hydrogen peroxide and acetic acid (vinegar). One issue with this solution is that you can't store it for very long. PAA breaks down quickly, so the solution is to use two spray bottles and mix it on the fly. It has an acrid odor resembling acetic acid, produced by ants.

If the water is cloudy, let it settle and filter it through a clean cloth, paper towel, or coffee filter first.

Disinfecting water using bleach is straightforward, depending on the volume you seek to treat.

1. Locate a clean dropper.
2. Suck up fresh liquid chlorine bleach stored at room temperatures for less than one year.

3. Use table 19 as a guide to decide the amount of bleach you should add to the water: for example, use 8 drops of 6 percent bleach, or 6 drops of 8.25 percent bleach, to each gallon of water. Double the amount of bleach if the water is cloudy, colored, or very cold.
4. Stir and let stand for 30 minutes. The water should have a slight chlorine odor. If it doesn't, repeat the dosage and let stand for another 15 minutes before use. If the chlorine taste is too strong, pour the water from one clean container to another and let it stand for a few minutes before use.

Table 19: Bleach Amounts for Purifying Small Amounts of Unclean Water

Volume of Water	Amount of 6% Bleach to Add*	Amount of 8.25% Bleach to Add*
1 quart/liter	2 drops	2 drops
1 gallon	8 drops	6 drops
2 gallons	16 drops (¼ teaspoon)	12 drops (⅛ teaspoon)
4 gallons	⅓ teaspoon	¼ teaspoon
8 gallons	⅔ teaspoon	½ teaspoon

**Bleach may contain 6 percent or 8.25 percent sodium hypochlorite. Check the label.*

WATER SYSTEMS

One of the key engineering feats of ancient Rome was their construction of extensive aqueducts channeling water in from remote provinces. Using a variety of construction materials such as brick, stone, and concrete, Romans perfected the gravity-fed system for water delivery starting in 312 BCE. About 600 years later they had eleven aqueducts to bring clean water to more than 1 million residents. The Romans used an ingenious system of sluices, sedimentation tanks, cisterns, and gates to settle contaminants, remove leaves (and fish), and ultimately deliver clean water from hundreds of miles away to the baths, estates, gardens, and fountains of the day.

Neolithic cultures in South America and Central America also engineered impressive water systems. A 2010 article at ScienceDaily.com describes how two Penn State researchers teamed up with an archaeologist and a hydrologist to report on a unique water feature found in the Mayan city of Palenque, Mexico. The team described a spring-fed conduit located on steep terrain. With an elevation drop of around 20 feet from the tunnel entrance to the lower outlet about 200 feet away, and a decrease in the opening from 10 square feet at the top to less than 1 square foot at the bottom, the effect would have been great water pressure.

A study by a University of Cincinnati team, published in 2020 in *Scientific Reports*, describes how 2,000 years ago the Maya of Tikal built "sophisticated water filters" using natural materials such as quartz and zeolite they gathered from several miles away. A multidisciplinary team of anthropologists, geographers, and biologists knew that zeolites, a crystalline compound commonly used today in kitty litter, is a natural molecular sieve.

UC geography professor and coauthor Nicholas Dunning found the likely source of the quartz and zeolite about ten years ago while conducting fieldwork in Guatemala. "It was an exposed, weathered volcanic tuff of quartz grains and zeolite. It was bleeding water at a good rate," he said. "Workers refilled their water bottles with it. It was locally famous for how clean and sweet the water was."

Any kind of waterborne disease outbreak is concerning, and city and county water treatment plants add a variety of chemicals to ensure clean, safe water. Basically, treatment plants use four steps to treat water for consumption:

1. **Coagulation and flocculation.** Add alum to clear suspended particles.
2. **Sedimentation.** Allow the floc to settle to the bottom of a tank or basin while pumps or gravity move the clear water to another step.
3. **Filtration.** Remove the sediments that may not combine with flocculants using a filter of sand, gravel, and charcoal. Small-mesh screens can further remove contaminants.
4. **Disinfection.** Add a form of chlorine to kill remaining biological contaminants such as parasites, bacteria, and viruses.

Additives such as chlorine and fluoride are now a fixture of modern systems. On January 25, 1945, Grand Rapids, Michigan, became the first community in the United States to fluoridate its drinking water to prevent tooth decay. Today, the addition of minute amounts of fluoride occurs in over 80 percent of the major US drinking supplies. Fluoridated water creates very low levels of fluoric acid in saliva, which reduces the rate at which tooth enamel decays and increases the rate at which it can reform in the early stages of cavities.

Home water systems using wells or cisterns lack fluoride, but toothpaste can make up most of the difference if you brush regularly. There is no easy hack for adding fluoride to your own homestead system; you can't just periodically add a pinch of ground-up fluorspar crystals or fluorite specimens, for example. You'd run the risk of adding too much, creating the strongly corrosive hydrofluoric acid, and doing more harm than good to your precious teeth.

Campers and Homesteaders

If you are setting up a homestead or extended camping trip, one of your first tasks is to establish a clean water source. The days of dipping water directly from the nearby stream are just about gone—even well water can be contaminated with bacteria or chemical traces, and animal or human waste is prevalent in just about every stream and river. Ponds and lakes are even more susceptible. In short, bacteria and viruses are everywhere. According to the CDC, these are the main concerns for waterborne contaminants:

1. **Giardia** is a tiny parasite that causes diarrhea. Areas with heavy traffic from humans or animals often have traces of solid waste in the water. If those animals or humans are already infected with giardia, the disease can easily spread.
2. **Legionella** is a bacteria that causes a lung infection similar to pneumonia and spreads typically from stagnant or standing water. Small droplets of contaminated water spread through the air and enter the lungs, where they flourish.
3. **Norovirus** is a virus that causes vomiting, nausea, stomach pain, and diarrhea and spreads easily through contaminated water.
4. **Shigella** is a bacteria that causes an infection that leads to bloody diarrhea, fever, and stomach cramps. It is extremely contagious upon contact with feces.
5. **Campylobacter** is another bacteria that leads to diarrhea and stomach issues. It lurks in tainted, raw poultry but also can spread through untreated drinking water.
6. **Salmonella** is a bacteria that hides in birds, turtles, and even plant-based brie products. It leads to diarrhea but can worsen to cause typhoid fever. It spreads in water, on untreated surfaces.
7. **Hepatitis** is a liver infection caused by the hepatitis virus, which spreads easily though close personal contact with blood or feces. It can cause fatigue, nausea, stomach pain, and jaundice.

8. **Cryptosporidium** is a microscopic parasite that causes diarrhea and is the leading cause of waterborne disease in the United States.
9. *E. coli* is a bacteria associated with common food poisoning; it travels in water exposed to spoiled or undercooked food, urine, and feces.

That's a daunting list of bugs to watch out for, and it gets worse. Cholera is a sometimes lethal bacterial disease causing diarrhea and dehydration due to contact with sewage. In 1817 the first recorded cholera pandemic originated in India near Calcutta, killing perhaps 250,000 people. In his book *Plagues in World History* (Rowman & Littlefield, 2011), John Aberth notes that another outbreak, beginning in 1829, killed an estimated 350,000 people worldwide; a third outbreak, beginning in 1846, killed at least 1 million people in Russia alone.

According to the National Park Service website entry "Cholera: A Trail Epidemic," pioneers embarking on the Oregon Trail regularly succumbed to the disease. The phrase "healthy in the morning, and dead by noon" was a somber warning. One traveler, George Tribble, who journeyed in 1852, recalled that the first 400 miles of the trek was "almost a solid graveyard"—one camp had seventy-one graves.

The odds of encountering one of these dangers is greater as you move down the mountain, but even the springs and upper sources of creeks and rivers can be contaminated. The best defense is to boil water at a full, rolling boil for at least 1 minute. Note that the dead parasites and microorganisms settle to the bottom and that boiling merely kills them. You should also consider a filter.

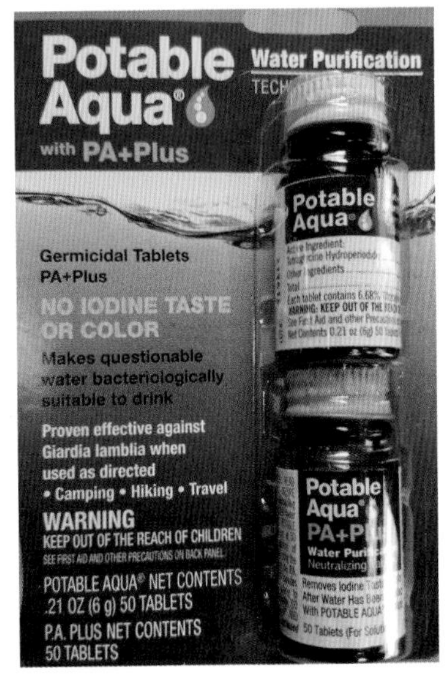

Make sure your water purification tablets kill *Giardia lamblia*.

Adding water purification tablets is a common camping method to treat suspect water sources. Most products rely on chlorine or iodine tablets. It usually takes 30 to 60 minutes to treat the water.

Another method for disinfecting water is through UV light. Small, lightweight systems such as pens that you stir in the water are effective. In a more dire situation, if you have a plastic bag, you can leave the water out in the sunlight for a few hours to days. Given enough time, the direct sunlight contains enough UV light to kill pathogens, but only in very small, shallow amounts.

A **water filter** removes waterborne protozoa and bacteria but not viruses, which are usually an order of magnitude or more smaller. A **water purifier** is designed to remove protozoa, bacteria, *and* viruses, providing a higher level of defense. Water bags and systems such as purification straws are also effective filtration devices, but most only remove particulates—only a reverse-osmosis system removes harmful bacteria.

The Life Straw is saving lives all over the world.

Purifiers offer more protection than filters, but two steps mean more work. If human traffic is relatively light where you are, the main threats are usually cryptosporidium and giardia and bacteria such as *E. coli* and salmonella.

In Case of Apocalypse

For urban and suburban residents, one of the most frustrating situations in a power outage is the gradual loss of clean water. Early in an emergency, move directly to the bathtub and fill it. That extra 50 gallons could end up being a lifesaver. If you're already in a "boil water" advisory, get out your bleach and eye dropper and calculate 6–8 drops of bleach per gallon of water. You might want to treat the bathtub as your indoor pond and only treat small amounts of water at a time in some kind of jug or pitcher, where you can more accurately calculate the water volume.

GEOLOGY HACKS FOR OUTDOOR TOILETS AND LATRINES

One of the key issues for homesteaders, campers, and survivalists is how to safely avoid issues with human waste. It's a gross subject, and there are far more detailed write-ups for plans and designs, ranging from pit latrines to elaborate composting structures with lighting and seating. From a geological perspective, there are two factors to consider: coverage and decomposition.

Coverage

The best answer for dealing with human waste has always been deep burial—especially for camps hosting lengthy stays or continual use. If you're interested, you can usually find plenty of write-ups about the history of latrines and toilets around World Toilet Day, November 19. According to WorldToiletDay.info, the United Nations estimates that some 4.2 billion people in the world do not have access to safely managed sanitation.

Much of the underdeveloped world uses slit-trench latrines—sometimes an embarrassingly open-air affair, with piled-up soil along a trench offering handy access to cover waste material. Adding mulch or healthy soil serves to quickly biodegrade material. (Adding poor soils such as clay isn't as much help.) The common approach for most hikers, campers, and survivalists is to dig a hole several inches deep and cover it afterward with topsoil. If you use common courtesy, you'll be fine. Keep groundwater pollution in mind, and stay far away from camp, water sources, and trails.

Decomposition

Their fact sheet on understanding soil microbes and nutrient recycling at Ohioline.osu.edu states: "There are more microbes in a teaspoon of soil than there are people on the Earth." That comes in handy when using soil for decomposition. As the fact sheet explains, actinomycetes are thread, or ray, bacteria that contribute mightily to soil health because they break down resistant materials, such as insect skeletons. Fungus populations require undisturbed or non-tilled soil to remain at healthy levels. Bacteria, actinomycetes, and protozoa are hardy and can tolerate more soil disturbance than fungal populations, so they dominate in tilled soils, while fungal and nematode populations tend to dominate in untilled or no-till soils.

Controlling Odor

One common method for controlling odor in latrines is to add lime—typically calcium hydroxide, with the chemical formula $Ca(OH)_2$. It's an odorless white powder that you can produce by adding water to quicklime.

Quicklime, known to chemists as calcium oxide, has the formula CaO. That's one calcium atom and one oxygen atom. You can create your own quicklime by burning limestone or seashells ($CaCO_3$) then quenching them in water. Now you know why another name for quicklime is burnt lime. It's a simple process—lay a large fire with plenty of wood and toss ample amounts of crushed limestone in multiple layers. Burn it and then dig out the cooked rocks and toss them into a water-filled pit, tub, small pool, or even a wheelbarrow. Let stand to evaporate.

Here's the magic when you apply heat:

$$CaCO_3 \rightarrow CaO + CO_2$$

See how that balances? There is one calcium ion, one carbon ion, and three oxygen items, but the heat rearranges them. Put another way, when you apply heat to simple limestone, one calcium carbonate molecule yields one molecule of carbon dioxide and one molecule of calcium oxide (quicklime). Obviously, you'd want to do this without breathing the fumes. Note that if the limestone has a high mixture of dolomite, a magnesium-rich variety, the process still works. The drawback for dolomite lime applies mostly when using it as a soil amendment—magnesium can build up quickly in the soil, and plants usually can't handle a lot of magnesium.

Table 20: Chemical Names and Properties in the Limestone Cycle

Chemical name	Other names	Formula	Common Source
Calcium carbonate	Calcite	$CaCO_3$	Limestone, shells
Calcium oxide	Quicklime, burnt lime	CaO	Heated calcite
Calcium hydroxide	Lime, slaked lime	$Ca(OH)_2$	Quenched calcium oxide

To convert quicklime to the slaked lime (calcium oxide) used to speed decomposition in your latrine, you toss the burnt lime into water. This reaction can be violent; it is exothermic, meaning the conversion from quicklime to calcium hydroxide gives off heat. In fact, it can appear to boil. Once the reaction has cooled off, you have a white, soupy mess. You can let it settle, pour off the clearest water, and dry over several days to a powdered form.

FuseSchool has an excellent video on YouTube for the limestone cycle. One of the YouTube videos produced by Primitive Technology shows how to use snail shells to create lime mortar, if limestone is in short supply. TransAmerica also has a video showing how to make quicklime using a traditional kiln.

GEOLOGY HACKS FOR WATER FILTRATION

If you're feeling pretty good about your water source and just want to remove most of the troublesome sediment, a clay vessel or 2-liter plastic bottle comes in handy, but a shirtsleeve also works. The method is simple:

1. Select a gallon-sized pot or container with two or three small holes in the bottom.
2. Add a coarse, pea-sized gravel layer on the lowest level, about 1 or 2 inches in thickness.
3. Add a thicker layer of clean, coarse sand, about three times the thickness of the coarse gravel layer.
4. Top off with a thinner layer of fine sand, about double the thickness of the pea gravel layer.

The goal is to allow the water to drip slowly but surely through the column. You may have difficulty finding two types of sand, but don't overthink it. Generally, if you have a bulky sand sample, you can swish it around and let it settle. The coarser sand will roll around in a different spot. Then you can scoop out the finer sand into another container, stopping when your fingers feel the difference in the grit. Otherwise, you can look for coarser sand behind or below boulders, in abandoned water channels away from the main water course, or at the top end of a gravel bar, where the heavier material falls out first.

The filter described above doesn't use charcoal, but if you have chunks available, you can use them near the bottom layer, before you place your pebbles. Charcoal offers another level of filtration, and it's a simple if somewhat messy process: Pull out the big pieces of charcoal from a recent campfire, place on a large rock, and grind with a smaller rock. You can use leaves, twigs, or moss to keep the charcoal from pouring out too fast.

At Survival.com there's a good DIY hack for building a charcoal filter. The advantage of using charcoal to filter out contaminants is that charcoal has more oxygen standing by, sitting alongside the carbon atoms of the burnt wood. Activated charcoal has been manufactured at extremely high temperature with quite a bit more available oxygen atoms, but regular burnt wood will work; it just takes longer—it's also cheaper. Just make sure water is dripping through, not pouring through, your filter. You can run the water through multiple times if it is still coming out murky.

Note that you should still boil (or chemically treat) the filtered water if you want to be completely safe from waterborne bacteria and viruses.

Variations on the filter include a tripod water filter, with sheets, shirts, bandanas, or other rigging. You can filter more water this way, but the methodology is the same: charcoal at the bottom, then sand and gravel, then finer sand. Pour in water from the top, and line up the seepage so that water travels from one station to the next, with a container to catch the final product at the very bottom.

Digging a Seep

A common method for creating potable water with few tools or accessories is to hand-dig a small seep next to a running stream. The idea here is to use the natural stratification of a gravel bar or streambed as your impromptu filter. By digging into the streambed, the theory is that you have a similar stratification to the common earth-filled filter, with a layer of big rocks followed by progressively smaller rocks in layers. The water tends to seep in from the bottom and, if slow enough, should be adequately filtered for emergency situations. Again, it's best to at least boil the water from a seep, but if you're high in the mountains, without the means for a fire and a pot to boil water in, a seep is often suitable.

If the first seep still seems a bit muddy, you can construct a second seep close by and double up the natural filtering effect. You can also use your shirt fabric to improvise a moderate filter to remove larger debris.

Speaking of your shirt, you can tie off the sleeve and fill with charcoal, gravel, and sand to create a fast filter. You don't even have to cut off the sleeve.

In arid desert regions, you can dig a seep in a dry ravine or gulch to locate water if plenty of water moves through periodically. Even in dry months, many streambeds with a large drainage above them have a thin green line of vegetation along one edge.

CHAPTER 7

ORIGINAL TOOLS AND WEAPONS

The average adult in good condition can last up to three weeks without food, but those last days are going to be gruesome. You won't have any energy to do much besides sleep, and you won't sleep well, dreaming of your favorite foods. After a week without significant calories, you will be light-headed, lethargic, make poor decisions, and generally suffer. To get food—especially protein—in most cases will require tools, so these topics are combined in this chapter. First, let's talk about geology hacks for food.

The easiest food to appreciate is food that requires very few calories—and little danger—to collect. Think nuts and berries, fruits and tubers, and maybe insects and earthworms. You'll need a digging stick; and to create a good digging stick, you'll need to apply some practical geology.

Sharpening a stick is fairly straightforward if you have a fire going. By alternately roasting the end and then sharpening it on one of the rocks in the fire ring, you can hone wood down to a nice, sharp point. If you select the right young sapling, such as a young ponderosa pine, you might also get a decent weapon suitable for taking down a large land mammal.

As weapons go, a sharp stick has limits. You can't use it as a throwing spear, because you can't hurl it fast enough to both fly straight and arrive with enough force to allow the pointy end to penetrate thick hide. That's why our ancient hunter-gatherer forebears quickly learned to turn quartz rocks into spearpoints and arrowheads.

TOOLS FOR SURVIVAL FOOD

For most survivalists, the snare is one of the most useful tools for catching small prey to fill the cooking pot. Usually fashioned from cordage and sticks or, in modern times, using wire, there is little to say about snares from a geological angle.

Under the "Rocks as Weapons and Tools" section we'll touch on throwing rocks, which require deadly accuracy and lots of work. Most humans past the age of forty are already starting to experience pain in their rotator cuff, the muscle set that controls movement of the arm in the shoulder. Thus throwing anything is a task for younger people. A potential variant will be slings, used by the ancients for hunting small game, and requiring surprising attention to detail in selecting the best stones.

Yet another ancient hunting device was the throwing stick, exemplified by the Australian boomerang, or the throwing club. There isn't much of geological interest there either, other than the hand axe or scraper used to fashion the wood.

Still, there is one ingenious method for hunting small game that has geological implications and still proves useful today—a heavy stone set with a hair trigger to flatten small game.

Deadfall Trap

Numerous websites and survival books offer tips for making a successful deadfall trap. PracticalSelfReliance.com, OutdoorLife.com, Instructables.com, and SelfRelianceOutfitters.com all feature instructions, and there are YouTube videos as well.

Note that according to the OutdoorLife.com article by Tim Macwelch, deadfall traps should only be used in a survival situation. They are illegal in most states unless you are specifically after rodents, in which case the law gets fuzzy state by state. In a 2017 report, BornFreeUSA.org passed out grades for all fifty states based on their animal trapping rules and regulations, which the group generally wants more of. Idaho and Alaska support trapping in general, while California strictly regulates trapping of all kinds. Iowa rated an "epic fail" because it has no prohibited animal traps, no restrictions against how trapped animals can be killed, no trapper education requirement, and no restriction on the number of traps, to name a few of the reasons for BornFree's ire.

In a survival situation, fines are the least of your concerns, and the concept of a deadfall trap is simple. You rig a sensitive trigger that holds bait and then balance a large, flat rock so that when the bait moves, the trigger releases and the rock slams down. Two main trigger variants are the "split stick" and the "Figure 4." The Figure 4 trigger is more complicated, using a series of notched sticks in the classic "4" configuration, with the rock resting on top.

The split stick approach uses three sticks—two cut sections about 1 inch in diameter, with a deep groove carved in both cut ends of the large stick. The third stick is about the width of a pencil, and longer. It will serve as the trigger, wedged into the two-part post, and with the baited trigger set in the grooves you carved. Here's a simple process. (Make sure you scrub your hands with a pungent natural scent such as conifer needles, mint, or wild onions before handling the raw materials.)

1. Find a flat surface on rock or hard-packed earth to prevent your prey from surviving the falling rock.
2. For the trap, select a large stone with a flat surface, about 20 inches long, 10 inches wide, and 5 inches thick. Ideally, according to Paiute Indian lore, the stone should be at least five times the weight of the animal you're hunting.
3. Select a thick stick about 10 inches long and 1 inch thick for the support post.
4. Saw the stick in half and trim the ends flat as well.
5. Select a smaller stick for the trigger, about 10 inches long and ¼ inch in diameter.
6. Carve deep grooves to allow the trigger to slide in when the grooves join up.
7. Test the depth of the grooves with the smaller stick, cutting it down if necessary.
8. Place the two halves of the support post together, hold in place, and balance the rock on the structure.
9. Bait the trigger with peanut butter, animal fat, insects, fish guts, or some other aromatic bait.
10. Insert the trigger stick into the carved grooves.
11. Test with a long stick.

Once the trap is set, you can scatter some attractant such as broth, wild onions, kitchen scraps, even flower petals, to hopefully mask your human scent and bring in curious game. Practice will allow you to settle on the best bait. Typically, you'll want something unique or exotic to attract interest, but more likely, especially in a survival situation, all you'll have to use is what's in the area.

A small Paiute-style deadfall trap, made with dogbane cordage PHOTO BY "YOURCELF." THIS FILE IS LICENSED UNDER THE CREATIVE COMMONS ATTRIBUTION-SHARE ALIKE 3.0 UNPORTED LICENSE, EN.WIKIPEDIA.ORG/WIKI/EN:CREATIVE_COMMONS.

If you don't have a pocketknife, you can build your system and skin your catch with a sharp rock flake. Smaller animals can be baked in the coals of a low fire in their furry jacket or encased in clay.

ROCKS AS WEAPONS AND TOOLS

At various sites, including Smithsonian Institution's humanorigins .si.edu, the general march of human evolution is explained in various stages, shown roughly in table 21.

Table 21: Human Origins and Tool Use Evolution

Time	Geologic Age	Hominid Species	Key Events
1 million years ago	Pleistocene	*Homo erectus*	*Homo sapiens*—"wise man"—emerges
2 million years ago			Earliest use of fire 1.5 million years BP
3 million years ago	Pliocene	*Australopithecus*	Stone tools ~ 3.4 million years BP
4 million years ago		*Ardipithecus*	Earliest bipedal stance
5 million years ago			
6 million years ago	Miocene	*Orrorin*	Hominids split from chimpanzees
7 million years ago		*Sahelanthropus*	
8 million years ago		*Oreopithecus*	
9 million years ago		*Ouranopithecus*	Chimpanzees split from gorillas
10 million years ago		*Nakalipithecus*	Early apes

Scientists previously dated the earliest stone toolmaking to about 2.6 million years ago, based on carved up bones found in Ethiopia. Then, in 2010, researchers from the California Academy of Sciences discovered evidence that our human ancestors butchered large mammals nearly a million years earlier than previously understood. Bones with evidence of cuts, scrapes, and smashes (to get at marrow) from the Dikika, Ethiopia, site sat in volcanic ash beds about 3.4 million years old.

Clearly, in a primitive or survival situation, rocks have many uses, as both weapons and tools. Due to their resistance to wear compared with more fragile artifacts, rocks show up in the archaeological record in many ways. What follows is a more or less chronological order for how rocks entered the tool kit of ancient practical geologists.

The Hammerstone

It's impossible to pinpoint the first humanoid use of a tool. It could have been a thick, hard tree branch or a large mammal thigh bone used as a club, or it could simply have been a rock held in a fist. Most likely it was whatever was at hand, but there's no doubt that the first stone tool was a hammerstone. Simply put, scientists define a hammerstone as an ancient stone tool used as a hammer, such as for chipping flint, processing food, or breaking up bones.

Any roundish, fist-sized river cobble will do, but some rocks tend to break apart faster. Sandstone or volcanic tuff doesn't last long; nor would such rocks deliver much energy when pulverizing long tibia and fibula bones for marrow. Volcanic rocks such as basalt, rhyolite, and andesite are reasonably hard and smooth, and readily available in the riverbeds of most African drainages. Metamorphic rocks such as gneiss, by contrast, are supremely hard—but sometimes retain sharp edges, even after quite a bit of tumbling. So, through process of elimination, we can imagine that basalt river cobbles were likely an easy source for the first human hammerstones.

There are plenty of potential hammerstones in this riverbed.

The Hand Axe

A hand axe is a simple tool that has been worked with some kind of craftsmanship, as opposed to a hammerstone, which may be unworked. The hand axe is therefore commonly referred to as the longest-used tool in human history. Archaeologists may refer to them as a biface, as they are two-sided, featuring two flaked faces. Usually knapped from flint or chert, hand axes usually have a pointed end and a rounded base and, as the name implies, were gripped by hand. Mostly they were used to dress game, chop wood, remove bark, or dig out roots and small mammal burrows. Their utility was undeniable; if they broke into pieces, the fragments could probably be used as slicing tools.

By about 1.76 million years ago, early humans began to make crude hand axes and other large cutting tools. Archaeologists refer to that early pattern of shaping round river cobbles into choppers as Oldowan, named for Olduvai Gorge, where the pattern was first noted. Oldowan tool kits include hammerstones, stone cores, and sharp stone flakes. Flint is less common in Africa than chert, so many of the tools are a form of chert or jasper.

The next pattern of construction is labeled Acheulean, for the type locale in France where the distinctive oval and pear-shaped hand axes first gained notice. This style of tool creation is attributed to *Homo erectus* and *Homo heidelbergensis*.

Acheulean tools were produced during the Lower Paleolithic era across Africa and much of West Asia, South Asia, East Asia, and Europe, and are typically found with *Homo erectus* remains. It is thought that Acheulean technologies first developed about 1.76 million years ago, derived from the more primitive Oldowan technology associated

Acheulean hand axes from Kent (from the Victoria County History of Kent, vol. 1, 312, published London, 1912, and now in the public domain)

with *Homo habilis*. The Acheulean culture lasted as late as 130,000 years ago. In Europe and Western Asia, early Neanderthals adopted Acheulean technology, transitioning to the Mousterian style by about 160,000 years ago. It isn't important to know all the different technology styles; just understand that experts can spot the differences immediately.

The term "hand axe" was first used by French archaeologist Louis Laurent Gabriel de Mortillet (1821–1898) when he suggested clarifying and reordering the Paleolithic. Previous researchers differentiated between prey animals, such as mammoths and reindeer. Mortillet used sites and artifacts in his 1869 work.

Due to their utility, hand axes are common in the archaeological record. Just about any rock that fits comfortably in your hand and can be worked to an edge can serve as a hand axe, but flint, chert, jasper, and obsidian all yield suitable tools and can be fashioned by cradling the rock and striking with a hammerstone.

The Scraper

In addition to crushing bones, hammerstones have another important use—fashioning sharp edges on chert, flint, agate, obsidian, jasper, or whatever member of the quartz family is available. Thus hammerstones helped to create the next likely tool—the scraper.

The blow used to flake off a useful sharp edge takes practice, but it's not a hard skill to master. The rounded edge of the hammerstone is a key; the effect of striking a second rock is to push a percussion build, or shock wave, through it. Most of the quartz derivatives are cryptocrystalline; their crystal nature is only vaguely revealed under a microscope. The result is the classic conchoidal (con-KOY-dee-al) fracture—from the Greek *konche* for "mussel shell." Various authorities describe the conchoidal fracture: smooth, curved surfaces, slightly concave, sometimes with concentric lines or undulations that strongly resemble seashells such as clams. This makes such rocks prime candidates for whacking into slices, sometimes called blanks, rather than fragments. Master crafters then carefully chip away at the edge to make the cutting side into a strong slicing tool.

Cutting meat from the bone is much easier than chewing it off or ripping it away. A sharp edge on the right rock can cut through cartilage, tendons, and sinews, quickly rendering game into dressed meat. In fact, there is evidence that ancient toolmakers had a keen eye for the rocks that yielded the best cutting edges. Tomos Proffitt, a researcher at the

University of Central London, coauthored a report in January 2020 for the *Journal of Royal Society Interface* in which he pointed to complex decisions involved in raw material selection.

Oldowan toolmakers used three main raw material groups for creating their stone tools—basalt, chert, and quartzite, all plentiful around Olduvai Gorge. Proffitt believed the ancient toolmakers chose different rocks depending on the type of tool needed. Chert and quartzite made for sharper edges, but basalt was more durable. Optimizing the resource choice ensured that tool efficiency and "ease of use" were the end result.

"The decision processes underlying raw material selection behaviours represents a major element of Palaeolithic research, and is often vital to interpreting the behaviour and cognitive capabilities of early hominins," the report stated. Simply stated, our ancestors were very smart rockhounds.

THROWING ROCKS

To bring down small game such as squirrels, rabbits, and birds, primitive hunters likely tried to use rocks early on. In *A Dynamical Analysis of the Suitability of Prehistoric Spheroids from the Cave of Hearths as Thrown Projectiles*, a 2016 paper by researchers led by A. D. Wilson and published in the journal *Scientific Reports*, the authors analyzed ball-shaped stone objects found in various African archaeological sites dating to 1.8 million years ago. They determined that 81 percent of the stones from the Cave of Hearths in South Africa were the size and shape suitable for inflicting "worthwhile damage" if hurled with force from up to 25 meters in distance.

In the report, "spheroids," as they called them, were acknowledged to be useful for multiple activities, such as to make stone tools, process plant material, break down bones, and resharpen grinding slabs. They proposed that these round objects were primarily cached as tools. Other research indicates that rock throwing probably predated the use of stone-tipped spears. The authors also reported that some of the throwing rocks were diabase—a common dark-colored igneous rock similar to basalt—but there was no local source for this rock. Therefore the stones must have been prized enough to carry for long distances. Other throwing rocks were made from quartzite, quartz, chert, and granite.

If you've ever casually tossed rocks from a beach into the water, you can understand what must have been going through the minds of prehistoric hunters. Some rocks are too light and don't carry their speed well. Other rocks are too heavy to throw far enough with accuracy. And some rocks aren't round enough to avoid curving off in an unpredictable direction. A smooth stone can slip, so just as modern baseballs have stitching to allow the pitcher to spin a curveball or slider, ancient throwers could have prized the ability to get a good grip. Analysis showed that most of the rocks recovered were in the range of 0.5 to 0.75 kilograms (1.1 to 1.7 pounds), and there was strong evidence that many rocks had been purposefully shaped to further improve their speed and accuracy.

Ancient Greek armies employed units called *petrobóloi* and *lithobóloi* ("stone throwers") as important combatants. Roman armies trained soldiers to throw rocks, up to 1 pound in weight. Later, stoning became a form of capital punishment in Jewish, Christian, and Muslim societies, and it still exists in some cultures.

Slings

Another ancient tool involving rocks as weapons was the humble sling. In ancient China it was known as "the whirlwind stone" (*piao shih*). Naturally, slings are not common in the archaeological record; they are usually made of a leather pouch cradle and cordage with finger loops at each end, woven from flax, hemp, or wool, and thus biodegradable. Slingers were repeatedly represented in ancient art. At AncientBattles.com, an article titled "Ancient Slingers" sets the date thanks to artwork from Çatalhöyük, in Turkey, circa 7000 BCE.

Archaeologists have discovered lead and clay "bullets" at numerous prehistoric battle sites, but unlike rounded rocks used for throwing, these manufactured projectiles for slings are shaped like miniature

Ancient Greek sling bullets with engravings, fourth century BCE. One side depicts a winged thunderbolt; the other, the Greek inscription ΔΕΞΑΙ ("take that") in high relief. PHOTO BY MARIE-LAN NGUYEN. GR 1842.7-28.550: COLLECTION OF THOMAS BURGON; PURCHASED, 1842; GR 1851.5-7.11: PURCHASED FROM P. J. GREEN, 1851 (RELEASED INTO THE PUBLIC DOMAIN).

footballs. A 2016 article in *Scientific American* by Tom Metcalfe, "Whistling Sling Bullets Were Roman Troops' Secret Weapon," described how drilled, oblong lead projectiles must have made a horrific noise when used against Scottish warriors some 1,800 years ago.

Perhaps the most famous example of a slinger is the biblical story of David bringing down the Philistine giant, Goliath, to the amazement of the assembled armies.

Clubs and War Hammers

Just about every major culture had a term for the war club. Aklys, baton, truncheon, blackjack, clava, cudgel, knobkerrie, nunchaku, shillelagh, slapjack, waddy—each refers to a wooden weapon capable of inflicting damage.

For the purposes of practical geology, we'll concentrate on war clubs fashioned from a single rock or that used worked stones set in a wooden handle. One of the most famous war clubs was the mere (MAIR-eh), fashioned from nephrite jade and shaped almost like a paddle, used by the Māori of New Zealand.

Stone-topped war clubs are often associated with North American Plains Indians, who fastened rocks to the end of a forked branch using pitch for glue and wetted leather or sinew that dried and tightened the connection. They mostly used chert or flint, sometimes substituting granite or basalt. The Maya used blunt jade axe-heads affixed to a wooden haft to break bones and shatter skulls, rather than cut or slice. Aztecs and Mayans both used obsidian, but due to its brittle nature, such weapons may have been more ceremonial than practical.

Adzes and Pickaxes

"Hafting" is the term for constructing tools that combine stone heads or points with a wooden handle or shaft. Hafting is considered to be an important innovation by early humans, as it involves complex planning and the use of several ingredients. Placing a sharpened stone on a haft provides much more power on the downstroke and greater reach. Humans are thought to have begun making hafted tools between 100,000 and 200,000 years ago.

While the use of rocks in war and combat is important, the more common use for stone-based tools among ancient humans was most certainly for agriculture and home improvements. One important tool was the adze. Typically used similar to a hoe, the adze was a versatile tool,

Adzes, Marshall and Yap Islands. Exhibit from the Pacific Collection, Peabody Museum, Harvard University, Cambridge, Massachusetts. IN THE PUBLIC DOMAIN

whether working with wood or tilling soil. Copper adzes replaced flint- or chert-tipped wooden adzes in Egypt's Predynastic period. Adzes are still used around homesteads for digging, chopping, and tilling.

Early occupants on the Big Island of Hawaii used the Mauna Kea Adz Quarry as a source for strong basalt to fashion tools. Located at an elevation of 12,000 feet, the complex features numerous trails, shrines, rock shelters, and petroglyphs, and is hailed as the largest primitive quarry in the world. The basalt there is fine-grained, free of cracks and inclusions for the most part, and thus ready to knap into large tools such as an adze. Archaeologists report that the Mauna Kea quarry rock has a distinctive chemical analysis, and tools from the quarry have been traced back from very distant sites in the Pacific Ocean.

Knives, Spears, and Arrowheads

The ancients soon learned to forgo fire-sharpened tips for their spears and arrows. They flaked out sharp chips of chert, flint, agate, jasper, obsidian, or whatever else they could fashion into a strong cutting edge. Like they did for war clubs, the ancients fashioned pitch as a glue with leather, sinew, or other cordage on a stout but lightweight piece of wood. Such weapons were important for both combat and hunting.

The concept is simple enough: Find a straight piece of wood, cut a notch in one end, and wedge in and/or tie on a sharp slice of rock. Use a big, heavy stick for spears; a smaller, thinner, and absolutely straight piece of wood for arrows; and bone, wood, or antler for knife handles.

The atlatl is a variation on spear-throwing that provides considerably more leverage. Basically, a short, notched stick fits into the butt of a throwing spear. The physics are straightforward—you preserve energy by changing the angle of your throwing motion, and you maintain energy by continually gripping the launcher, rather than battling the friction of releasing the spear shaft from your fingers. The increased velocity can mean the difference between piercing the thick hide of a woolly mammoth or watching the projectile bounce off harmlessly.

Over time, the ancients learned which rocks held up best under stress, which were easiest to work with, and how best to fashion grooves for tying the rock in place, no matter the tool. Before discussing the art of knapping in detail, there is one ancient expert we can "ask" to get an idea of a typical prehistoric tool kit.

Ötzi the Iceman

In 1991 hikers discovered what was eventually determined to be a 5,300-year-old mummified human form emerging from the ice of a melting glacier in northern Italy's Ötztal Alps. The mummy soon became an archaeological sensation, research subject, and museum display. From the available evidence, the man who came to be named Ötzi (pronounced "urt-zee" with only a faint "r") was crossing what is now the Austria-Italy border when he was killed. An X-ray scan revealed a flint arrowhead in his left shoulder, which severed the subclavian artery, so he bled to death quickly.

Research posted at iceman.it/en/the-mummy/#death indicates that the Iceman may have been involved in hand-to-hand combat in the days leading up to his death, as he had a deep cut on his right hand. Pollen and maple leaves in his birch bark container indicate that he died in early summer. His head wound suggests that he may have fallen off a cliff or into a crevasse after being struck by the arrow; that would explain why his tool kit and valuable copper axe remained in his possession.

(a) The oldest axe found complete with copper blade, hide strips, birch tar, and handle made of yew wood; it has been carefully dated by radiocarbon methods (figure from www.iceman.it, modified). (b) Casting defects and deformation in the talon of the copper blade. The microsample was extracted from the major cavity. G. ARTIOLI ET AL. (2017) "LONG-DISTANCE CONNECTIONS IN THE COPPER AGE: NEW EVIDENCE FROM THE ALPINE ICEMAN'S COPPER AXE," *PLOS ONE* 12(7): E0179263; DOI.ORG/10.1371/JOURNAL.PONE.0179263.

The copper axe is especially noteworthy; it places him in the Chalcolithic period, or Copper Age. Humans had thus emerged from the Stone Age and were on the path to metallurgical progress. The website has a lengthy description of the prized weapon: "The copper axe found with the Iceman was preserved intact and is the only one of its kind in the world. The blade consists of 99.7% pure copper and is trapezoidal in shape. The knee haft is made from yew and is approx. 60 cm long. The copper blade is fixed into the forked shaft of the haft with birch tar and is tightly bound with leather straps to hold it firmly in place. The blade was cast in a mold, cooled and then compressed by hammering. Signs of wear show that the axe had been frequently used and therefore had to be re-sharpened."

The rest of the Iceman's kit is particularly interesting for a practical geologist, which Ötzi definitely was. He carried a chert dagger blade, two completed arrowheads, three un-hafted chert tools, and a retoucher for rebuilding the sharp edge on chert tools. Research findings published in 2018 in the open-access journal *PLOS ONE* revealed that the material for his chert-bladed knife came from three different areas, up to 40 miles from where he died.

Ötzi also carried a small, sharp flake useful for making cordage by cutting reeds. His kit included an oval-shaped stone called an end-scraper, which had a working edge useful for processing hides and fashioning bows and arrows. He also packed a borer, a rock with a long, sharp end used to make holes in leather or wood.

The axe-head is not local. Despite the fact that copper ore sources in the Alps are known to have been exploited at the time, a study indicated that the copper in the axe came from southern Tuscany. The smith forged the axe-head from almost pure copper, using a combination of casting, cold forging, polishing, and sharpening.

Also among Ötzi's possessions was a type of tinder fungus, included with his fire-starting materials. As discussed earlier, catching the spark from a fire starter is a challenge, but for Ötzi it must have been fairly routine with his dried fungus. In North America, Chaga mushrooms, which grow on birch trees, can be ground into a powder. King Alfred cake and horse hoof fungi also work when dried.

ALL ABOUT KNAPPING

Scientists have begun pushing back on the notion that tool use is all that separates humans from other animals. Macaque monkeys at the shores of some Thailand islands have been so successful at using stones to break apart mollusks that the local shellfish population is threatened. Dennis O'Neill notes on his website (www2.palomar.edu) that chimpanzees are also occasionally observed as practical geologists: "Some chimpanzee communities are known to use stone and wood as hammers to crack nuts and as crude ineffective weapons in hunting small animals, including monkeys. However, they rarely shape their tools in a systematic way to increase efficiency."

That's what knapping is all about—increasing efficiency, especially in cutting. The term "knapping" comes from the old Germanic *knop*, meaning "to strike, shape, or work." Most of the tools used during the Stone Age were simple and efficient, used for dressing game and preparing plants. Eventually the art and science of fashioning efficient tools spread far and wide.

Ancient knappers would be unlikely to bring a heavy boulder home from a far-off quarry; instead they'd prep out pieces called blanks or cores in what's called hard-hammer percussion flaking. They'd be looking for large, solid pieces without fractures and pits. Using a hammerstone, a worker would whack the edge of a larger rock to spall off slabs, select the best ones, and bring them home for more careful work during the winter

A fine chunk of quartz material ready for more work back in camp

months. The technique takes practice, not only in applying the blow at the perfect spot but also in learning which basalt, quartzite, or gneiss hammerstones allow the most control. A hammerstone used to create preformed blanks has to hold up over time and allow for consistent power.

The perfect preform will be thin at the edges, fat in the middle, and have a good thick base. It could probably be hafted and used for a weapon—once. The issue is that the thin, sharp edge won't last very long. By tapering down the angle back toward the thicker middle, you add strength to the cutting edge and it will last longer.

To set the stronger edge, knappers use a copper "bopper," or billet, to take off larger fragments, then use a copper spike or deer antler point for finer work or, if damaged by use, to retouch the edge. Ötzi's retoucher (see sidebar) was a wooden block with a deer antler fragment attached.

Quartzite is tough, but it takes a lot more effort to fashion into a tool because the mineral grains aren't small and even. Quartzite often has a sugary appearance, with a crumbly rind on the outside; it usually won't flake off in a consistent manner, so it's mostly useful for scrapers.

Chert, flint, agate, chalcedony, jasper, and obsidian are the typical rocks knapped into sharp cutting tools. To understand what's going on inside the rock when struck, it's important to know about these minerals on an atomic level. You're probably familiar with the concept of a crystal lattice—the way a mineral's atoms line up.

You can't use crystals to make an arrowhead because they tend to cleave along crystal planes. We talked about cleaving earlier—think of mica, which cleaves into thin sheets. The best rocks for knapping don't

Modern tool kit for knappers (from bottom left): abrader, pad, large bopper, copper-tipped bopper, worked fragments, retoucher plus extra copper nail

cleave at all; they fracture in a characteristic pattern. If you've ever seen what happens when a kid shoots a BB gun at a glass window, you know that characteristic cone-like fracture. It's called a conchoidal fracture, as the shock wave moves from the point of impact outward in predictable, radiating circles. Knappers have a feel for this destruction and anticipate the physics involved to determine their strikes at the platform and control that fracture.

Knapping tool kits include the following:

1. Large "bopper" for creating blanks
2. Smaller copper bopper used for percussion flaking
3. Copper nail, for finer pressure flaking
4. Abrader, to rough up razor-sharp edges so you have a better platform to strike
5. Leather pad, so you don't tear up your jeans

Flintknappingtools.com has all these items for sale, but of course there are primitive substitutes. A deer antler, chunk of quartzite, and bark will substitute for the modern tools.

At the University of Texas anthropology department's website, there is a scientific study of three different knapping techniques:

1. **Hard hammer percussion** is the earliest and most basic flint-knapping technique. This produces finished but quite simple sharp-edged tools, sometimes called blanks; but more commonly they serve as the starting point. The hammer blow on the "platform," or flat edge, comes in as applied force at less than a 90-degree angle delivered at the point of percussion. The swelled part of the flake is the bulb of percussion. You might get occasional "erailleur scars," or chips, also knocked off the bulb, and there may be ripples.
2. **Soft hammer percussion** produces flakes from the blank and produces a diffuse bulb of percussion, usually from a very small platform. The strike angle is usually well beyond 90 degrees. Tools include antler tips, soft copper nails, or even a piece of hardwood. The goal is to produce thin, flat flakes. This allows workers to fashion both sides of the blank.
3. **Pressure flaking** produces small flakes and again uses soft copper, bone, antler, or wood. The angle is again well beyond 90 degrees. The result is much smaller flakes, usually thin and fragile, and often an abrader helps provide more of a platform to pressure flake. The goal is to shorten and toughen the angle so the edge isn't too brittle.

Ishi of the Yahi Tribe

American interest in knapping can partially be traced back to the study of a California Native American called Ishi, who lived in the early twentieth century. Rescued from starvation in 1911, he was celebrated as the "last wild Indian" and soon was sent to the University of California, Berkeley, where he worked as a janitor while battling various twentieth-century diseases. Before he died, he taught researchers what he knew about the traditional methods for making stone tools using his knowledge of knapping techniques. The much-publicized findings started a small craze in knapping among archaeologists and prehistorians, and today many groups devoted to knapping exist in the United States and around the world.

Even today, some of the best tools for procedures such as heart surgery are precision obsidian scalpels. Costing as much as ten times what a thicker steel blade costs, these delicate tools make an incision that's as much as 500 times cleaner than steel and therefore heals much faster. Under a high-resolution microscope, obsidian blade edges appear smooth and straight, while a steel scalpel resembles a saw blade.

HEATING ROCKS

At some point in the distant past, someone somewhere learned that heating certain rocks, especially flint, helps tremendously before knapping. The technique is simple:

1. Spread the raw flint out in a pre-dug pit.
2. Cover evenly with about 3 inches of sand.
3. Burn the fire at moderate heat—no bonfires!—between 500°F and 600°F for 5 to 6 hours.
4. Cool the flint for a couple of days.

The resulting flint becomes more uniform, and flaking results in harder, thinner waste. There are challenges of course; the center of the fire is always hotter, and some rocks will probably shatter. They might change color as well, but that's usually a good sign. All cherts and flints benefit from heat-treating before knapping, but obsidian and dacite don't need such treatment.

At this point we've covered most of the primitive skills that relate to geology, but you can only stay in the Stone Age for so long.

CHAPTER 8

METAL TOOLS

If there is one advance that marks significant human achievement, it's our ability to harness all the metals and convert them from rocks into tools. Table 22 shows the rough timeline for the key advances, but note that discoveries are still moving these dates around.

Table 22: Time Periods for Use of Key Metals

Age	Time Period
Iron Age	1,200 years—present
Bronze Age	3,300 years—1,200 years BP
Copper Age	4,500 years—3,500 years BP
Stone Age	3.4 million—4,500 years BP

These ages are approximations because some regions were more or less advanced than others.

THE COPPER AGE

Earlier we stated that the original center of metallurgy was likely the Iran highlands, since native copper nuggets were fairly common there. Excavators have found evidence of copper crafting in Iran, Mesopotamia, and Egypt probably as early as 4000 BCE.

During the third millennium BCE, the legendary city of Troy hosted numerous metal craftsmen, merchants, and traders, and their knowledge spread throughout the Mediterranean region. Since about 2000 BCE, copper crafting was widely diffused in Europe, according to an older but eminently readable article at Rameria.com, collected from various sources by a prominent Italian coppersmithing family. Workers soon learned to extract copper from its ores of cuprite, enargite, chalcocite, and bornite, among others.

Blacksmiths next discovered that metal hardens when hammered for long periods of time but can return to its softer state by reheating, without any loss of shape. Hammering, heating, and cooling became the main method used for creating copper products.

THE BRONZE AGE

Metallurgists seeking to add strength to pure copper eventually combined it with small amounts of tin to form bronze. From 3300 to 1200 BCE, copper and tin (with some arsenic) in the form of bronze was the dominant metal used for tools and weaponry. A big advantage of bronze was the ability to pour the liquid mixture into molds for casting, which enabled mass production.

Reasons for the collapse of Bronze Age kingdoms are unclear. Severe drought, earthquakes, political unrest, invasion by nomadic tribes, or all of the above may apply. Shortages of tin could also be a factor.

Ancient Greece

Notable historians such as Plato, Aristotle, Xenophon, and Pliny all wrote about ancient Greek mining and metallurgy. They described the technology that allowed miners to dig successively deeper mines and the increasingly sophisticated furnaces ancient technologists used to treat copper ores.

The Greeks devised a large smelter using charcoal as a reducing agent, with internal walls lined with fire clay. Temperatures reached from 1,000°C to 1,200°C (1,832°F to 2,192°F) with the aid of hand blowers to pump in air.

Ancient China

The Arkenstone (irocks.com) notes that the ancient Chinese began prospecting for useful minerals as much as 6,000 years ago, and were probably mining for metals 5,000 years ago. Researchers at Tonglushan discovered a 3,000-year-old copper mine with nearby smelting facilities. The Yin ruins at Anyang in Henan province yielded a sacrificial vessel known as the Houmuwu ding (formerly the Simuwu ding) that was especially noteworthy: It "contains 84.77% copper, 11.64% tin and 2.79% lead, which is very close to the composition of the bronze alloy having the highest hardness known to the modern metallurgical industry."

The Chinese learned to identify the best combinations of minerals that told them they were working valuable ore, and they determined which plants might indicate rich gold ores in the rocks below the soil layer.

THE IRON AGE

The Iron Age began around 1200 BCE in the Mediterranean region as the Indian, Greek, and Persian empires collapsed. The cause for the collapse of these Bronze Age kingdoms remains unclear. Archaeological evidence suggests that a succession of severe droughts in the eastern Mediterranean region over a 150-year period from 1250 to 1100 BCE likely figured prominently in the collapse. Earthquakes, famine, sociopolitical unrest, and invasion by nomadic tribes may also have played a role.

Some experts believe that a disruption in trade routes around this time may have caused shortages of the copper or tin used to make bronze. As a result, metalsmiths may have turned to iron as an alternative.

Copper has a melting point of 1,085°C (1,985°F), while tin melts at 232°C (450°F). Iron, by contrast, melts at 1,538°C (2,800°F), which required advances in furnace and kiln technology. Since iron is a frequent impurity in copper, experts believe the ancients mastered the technology to forge iron as their knowledge of making bronze increased.

The progression from simple iron to complex tools is fairly straightforward. Metallurgists originally smelted iron in furnaces called bloomeries. They used leather bellows to force air through a pile of iron ore and charcoal, relying on the carbon monoxide released from the charcoal to reduce iron oxide to metallic iron. Because the temperatures were still relatively low, the metal blob at the bottom of the furnace was a spongy mass called a bloom. Workers then hammered the bloom and folded and beat it some more to get rid of the accompanying slag, eventually producing wrought iron.

Wrought iron is not hard enough for effective tools or durable weapons. Metallurgists figured out that they needed to get more carbon into the sponge and developed the process of carburization, which involved heating the wrought iron in a bed of charcoal then quenching the red-hot sponge in water or oil. This rapid cooling caused the outer edges of the mass to convert to an alloy of iron and iron carbides: steel.

IRON METEORITES

There are numerous examples from history of iron meteorites finding their way into tools, jewelry, and weapons. "Iron from the Sky: Meteorites and Ancient Culture," an article at open.edu by Dr. Diane Johnson, describes one such event. In 1911 archaeologists discovered nine small iron beads in the burial goods of a man entombed at Gerzeh in Egypt as long ago as 3400 BCE.

Early researchers determined there was considerable nickel mixed with the iron, a telltale sign of extraterrestrial origin, as iron ores on Earth do not contain abundant nickel. In 2013 researchers at the Manchester Museum used a host of instruments and confidently concluded that the beads were definitely not terrestrial iron. "The . . . bead microstructures observed and their composition were consistent with that of an iron meteorite that had been worked into a small thin sheet and bent into a tube-shaped bead," Johnson wrote.

In addition, Johnson notes, a new word for iron appears in Egyptian texts around 1295 BCE: *Bia-n-pt*, which translates to "iron from the sky." She surmises that there may have been a major event around that time, such as a dramatic fireball across the sky. In 2008 a large impact crater was discovered in southern Egypt; the crater is less than 5,000 years old and may be the source of those objects.

Archaeometallurgist Albert Jambon, of the Pierre and Marie Curie University in Paris, was described in a 2017 article at LiveScience.com as using a nondestructive portable X-ray fluorescence (XRF) analyzer to scan some of the world's most ancient iron objects at museums around the world.

Jambon has identified meteoric origins for various iron objects, such as an iron dagger from Alaca Höyük in Turkey, dated to about 2500 BCE, and a ceremonial iron axe-head found at Bronze Age port city of Ugarit in Syria that dates from about 1400 BCE. He also verified an iron pendant found at Umm el-Marra in Syria that dates from around 2500 BCE and a set of iron axes from Shang dynasty China that have been dated to around 1400 BCE.

King Tut

Three objects of meteoritic iron were discovered in the tomb of Tutankhamen, who died in 1337 BCE. Johnson says the artifacts were described in temple inventories as "stars." There was a miniature headrest, an amulet attached to a golden bracelet, and a dagger blade with gold haft. Only the dagger blade shows expert craftsmanship, leading Johnson to suppose it may have been made outside Egypt.

Campo del Cielo

In Argentina in 1557, the Spanish governor of a northern province learned that the locals collected and beat into tools the iron masses they said "fell from the sky." Guessing there was simply a rich iron ore deposit nearby, he initiated a search; soldiers discovered a large iron mass sticking out of the ground, plus numerous smaller pieces scattered about the area,. The governor's report was ignored, but in 1774 the site was revisited. In 1783, soldiers used explosives to clear away the soil, discovering that they were looking at a single large object.

They named the area Campo del Cielo, for "field of the sky," but grew discouraged that they hadn't found a larger deposit than the single 15-ton boulder. The subsequent report was again ignored, but samples sent to the Royal Society of London revealed a composition of 90 percent iron and 10 percent nickel and their true origin was clear—they were meteorites.

The largest two fragments, the 30.8-ton *Gancedo* and 28.8-ton *El Chaco*, are two of the heaviest single-piece meteorites known on Earth. They rank with the 60-ton Hoba meteorite from Namibia and a 31-ton fragment of the Cape York meteorite, found in Greenland.

A pair of smaller meteorite fragments, ready to fashion into iron tools

The Hadschar al Aswad

There is one more interesting meteorite story before moving on. The sacred black stone known as the Hadschar al Aswad is the venerated stone inside the Kaaba, the cubic building in Mecca, Saudi Arabia. All devout Muslims must journey at least once in their lifetimes to the Kaaba. Since the stone has never been tested using modern instruments, there is considerable uncertainty about it. Some believe it is a baetyl, a true meteorite given to Abraham by the archangel Gabriel. Others think it might be a large "Wabar pearl," an impact glass of huge size.

FORGING YOUR OWN IRON

There are many YouTube videos describing how to forge your own iron from iron ore. As discussed earlier, you'll need a lot of charcoal, a furnace or kiln with a bellows, several tools, and a source of iron-rich rock. It's not a simple process, but it hasn't changed much over the years.

Find Black Sands

Fortunately there is a fairly straightforward hack for collecting iron ore. As anybody who has ever panned gold knows, the bottom of the pan after you reduce a bucket of pay dirt down to a teaspoon of concentrates is usually rich with black sands.

The beach at California's Fort Funston contains loads of magnetic black sands.

The black sands are actually tiny minerals of magnetite, ilmenite, and monazite, in addition to platinum and platinum group elements (PGEs). This group includes platinum, palladium, rhodium, ruthenium, iridium, and osmium. They have similar properties and are fairly valuable in large quantities, so some gold prospectors save all their black sands for a day when they have enough to smelt.

For you to collect enough black sand to forge an iron tool, you'll need helpers; a sluice, rocker box, or dredge will also come in handy. Another trick is to find an ocean beach known for black sands, such as south of the mouth of the Columbia River or near old Fort Funston, south of San Francisco. You can drag a large, powerful magnet housed in a plastic container, load it up, pop the magnet out of the tub over a plastic sack, and easily collect pounds of black sands in an hour at those locations.

Build a Furnace

You'll need a sizable furnace, called a bloomer, to heat up the iron ore. There are a number of plans available; some use wood fuel, but others use propane. At theCrucible.org they used 350 pounds of clay to form a large belly at the bottom of a 4-foot chimney. Some of the YouTube videos showcasing primitive skills depict survivalists creating multiple kilns out of clay-rich mud, often recycling their previous kiln material.

Add a Reducing Agent

The most common reducing agent is charcoal, which unfortunately comes in large chunks that you have to grind to a powder. Then mix the powdered charcoal and the black sands in a 1:1 ratio.

Charge the Furnace

This step involves bringing the furnace up to temperature, usually with flames coming out the top of the chimney.

Smelt the Iron

The heat of the fire and the carbon from unburnt charcoal combine to reduce the iron oxide, removing the oxygen from the iron ore and leaving pure iron. You have to control the temperature and ratio of fresh charcoal; otherwise, the iron absorbs too much carbon. At the base of the furnace, the carbon is so desperate for oxygen, it will pull it from your iron ore. Eventually, the iron particles fall to the bottom and form a molten blob, called the bloom. It's a large chunk of glowing metal.

Release the Slag

The slag, or molten silicates, is like a messy liquid glass. It is a residual slag that doesn't form into the sponge, so it pools up at the bottom of the furnace. At some point you'll have to pierce the clay furnace with a solid poker so the slag flows out of the bottom.

Retrieve the Bloom

The next step is to tear apart the furnace and reveal the glowing lump, or bloom. It will be fairly obvious—it's orange-hot, but it doesn't want to flow. You can liberate it from the furnace pieces and remaining charcoal.

Hammer Out the Slag

Using tongs, a hammer, and an anvil—plus safety equipment such as heavy aprons, eye guards, gloves, long sleeves, and whatever else you can think of—you beat the bloom with a heavy hammer, sending sparks flying, which are more slag you don't want.

An excellent blog post at AnitaChowdry.wordpress.com covers the basics of bloomery. The post describes a pair of smeltings—one using 60 kilograms of black sands and the other using siderite, or iron carbonate ($FeCO_3$). Chowdry reports that while the black sands produced high-carbon steel, the siderite produced wrought iron.

PRIMITIVE SKILLS VIDEOS

On the Survival Skills Primitive YouTube channel, two Southeast Asian brothers apparently show how to enter the Bronze Age in their video at youtube.com/watch?v=6LEAgMBzCTY. They collected rocks from a nearby river that appear to be rich in copper, perhaps zinc or tin, and perhaps arsenic. They then crushed and smelt the ore in waist-high kilns made from clay-rich mud, constructed with bellows using bamboo. They added charcoal to the ground-up rock powder, poured the mix into the top of the kiln with more charcoal, and heated it up to a glowing temperature.

After a while, they tore open the bottom of the kiln while it was still hot to fish out small, glowing metal nuggets from the bed of coals. They then cooled off the rest of the coals with water and plucked out the remaining bits of bronze. From there they used a porcelain crucible to melt the bronze nuggets into a molten mix within a smaller clay kiln, again using a bellows and loads of charcoal. Meanwhile, they created a knife mold in wet sand and poured in the molten bronze. Later they made a hatchet head and a chisel using the same process. From there it was a simple step to use rocks as sharpening stones to set the edge and polish the tools.

SHARPENING STONES

Now that we have discussed metal tools, it's time to show how geology impacts tool maintenance. To keep an edge and maintain a blade requires a very humble tool: the sharpening stone.

As we saw for ancient fire starters, the geology behind sharpening stones goes back to the Mohs hardness scale. Iron and steel rarely rise above 7 on the Mohs scale, so many quartz-rich stones can hone a metal, especially at the thin blade edge. Sharpening stones begin at 7 on the Mohs scale and reach all the way to a 10 in the form of diamond stones.

All sharpening stones abrade or cut the metal they are applied to. In general, the coarser the stone, the faster it cuts, but coarse stones don't yield good edges. You get a better edge with finer stones. The concept is similar to sandpaper—rougher grits get more done, but finer grits give more control. So if a blade edge is nicked or extremely dull, you'd start in with a coarser stone or even a metal file. You might use a medium stone as an intermediate step. Then you'd finish with the finest stone, which removes metal at the slowest rate.

The three most common classes of sharpening stones are oil stones, water stones, and diamond stones. Each of these stones has its own advantages that can help users achieve their sharpening goals.

Tools and oils for sharpening steel blades, including a file

Oil Stones

Oil stones are the familiar tools most of us grew up with. There are a couple of common oil stones. The Arkansas stone is made from novaculite, a sedimentary rock metamorphosed to chert from Devonian to Mississippian beds in the Ouachita Mountains of Arkansas, Oklahoma, and Texas in the United States, and also known in Japan and the Middle East. Novaculite has a hardness of 7 on the Mohs scale, and there are wide variations in color, toughness, and use for coarse, medium, and fine. Workers use a whetstone oil to carry off the fine metal powder ground off a blade.

There are man-made aluminum oxide substitutes in all grades and colors that are sometimes called India stones. Crystolon stones, another artificial whetstone, are made from silicon carbide. In general, all oil stones are the slowest at removing metal.

You can use just about any mineral or vegetable oil, or even water, with an oil stone. The honing oil is a lubricant that prevents small shavings from building up. The small pieces of metal have to get carried away from the stone instead of clogging the pores, which would ultimately decrease your stone's ability to cut metal. Note that plant-based oils can go rancid, so be sure to clean up after each session.

Water Stones

Water stones are easiest to work with—rather than requiring oil, a glob of spit will do the job. Few natural water stones find favor nowadays, as they are both rare and expensive. Synthetic water stones are generally made of aluminum oxide. They cut faster in a paradoxical way. Because they are actually a bit soft, you constantly remove tiny bits of the whetstone, but this actually exposes fresh material that's able to cut faster. That softness makes water stones prone to uneven wear, and they may need to be flattened (ground down uniformly) to bring them back to an effective tool.

Diamond Stones

Diamond stones really do have industrial diamonds embedded in their surface, and they cut the fastest of all materials, as you would expect from a tool that sits at the top of the Mohs scale. These stones are so effective, they also grind down other whetstones to restore their flat surface.

CHAPTER 9
MEDICINE FROM ROCKS

Medical geology is a specialized field in the earth sciences that investigates how rocks, minerals, and other geological factors impact us. Often, the news is bad: Volcanic ash, dust, and related "natural pollution" is dangerous for your lungs and skin, for example. Still, we also need trace amounts of elements and minerals along with our vitamins, and the sources for such components often come from geology.

CRYSTAL HEALING

In her book *ROCKS HEAL!: The Science of Rock-Medicine* (Bowers-Moser, 2015), Sela Weidemann goes into extensive detail about crystal and stone influences on the human body, arguing that "minerals are precision chemical instruments" that can be applied in the same context as allopathic and homeopathic remedies. She bases her theories on interpretations of physics and chemistry, and there has long been a field of healing that emphasizes crystal vibrations and other properties for good health.

Modern practitioners argue that their skills and influences trace back to ancient societies that believed in the power of crystals. And there is evidence that 6,000 years ago, the Sumerians practiced some form of medical geology. The Egyptians apparently believed that gemstones such as carnelian, turquoise, and lapis lazuli warded off illness and general "negative energy." The Chinese concept of Chi, or qi, is said to connect the physical and spiritual worlds, and the Buddhist idea of chakras—the vortices of life-energy—are another example of human energy manipulated by the proper amulets, talismans, and potions.

Attempts to verify these claims have never shown statistically significant improvement over a placebo, and many regard the ideas as a pseudoscience. Still, there are entire books on the subject, many at least partly based on ancient texts, believing, for example, that garnet balances energy and fosters creativity, or malachite clears obstructions.

On the other hand, never discount the idea of having a cool rock in your pocket. You can think of it as a good luck charm or simply a reminder of a productive rockhounding trip, and that alone may be enough to elevate your mood. There is little downside, after all, save for a sharp crystal tearing out some threads or a shiny pebble banging around in a washing machine or dryer.

Plus, the health benefits of getting out into the field and looking for new specimens are well known. Anglers like to say that a miserable day spent fruitlessly chasing an elusive trout is better than any day in the office. Rockhounds feel the same way about chasing down pretty rocks. Lately, an entire industry has sprung up around "forest baths," luring city dwellers into the cool, dark forest by extolling the virtues of simply slowing down and walking in the woods. Geo-tourism has also recently gained ground, with vacationers planning trips around erupting volcanoes in Hawaii or Iceland or touring ancient sites such as Stonehenge, Machu Picchu, or Easter Island.

Fresh air has more oxygen than household air, and getting out of the house and away from the city can soothe your brain thanks to less stress-inducing noise. Focusing on which rocks to pick up and which ones to leave right there (so-called leaverite) helps settle the mind, and it's hard not to smile when you find good material. All that adds up to a net plus—on that basis alone, communing with the geology spirits can be a beneficial and therapeutic activity.

Getting out and searching for rocks at low tide at the beach is great therapy.

THE ELEMENTS AND MEDICINE

There were a number of ways ancient healers took advantage of natural resources to medicate patients. Baking soda and Epsom salts still have a variety of uses, and they are quite inexpensive as well. The use of common metals in drugs can be traced back to ancient times. Gold-based medicines were being used in China and the Middle East as far back as 3,500 years ago. Any list of elements, minerals, and rocks used in pharmaceuticals and health care products will hardly be exhaustive, and nothing here should be taken as medical advice. The usages and notes for various elements listed below are excerpted primarily from the US National Library of Medicine and the National Institutes of Health (ncbi.nlm.nih.gov).

Arsenic: Arsenic and arsenic compounds have been produced and used commercially for centuries. Arsenic was used in some medicinal applications until the 1970s, when it fell out of favor; the United States banned it in 2003. Inorganic arsenic was used to treat leukemia, psoriasis, and chronic bronchial asthma, and organic arsenic was used in antibiotics for the treatment of spirochetal and protozoal disease.

Barium: Barium sulfate is used to help diagnose certain disorders of the esophagus, stomach, or intestines. It is an inorganic compound with the chemical formula $BaSO_4$. It works by coating the inside of the esophagus, stomach, and intestines, which allows those areas to be seen more clearly on a CT scan or other radiologic (X-ray) examination. Barium sulfate has very low toxicity due to its insolubility, and was also used in enemas.

Bismuth: Bismuth comes from the German word *wismuth*, for "white mass"; it was one of the first ten metals discovered, and over-the-counter medicines today contain bismuth subsalicylate (brand name Pepto-Bismol), used to treat nausea and diarrhea. If you enjoy chemistry experiments, look up the many YouTube videos that show how to recover bismuth from stomach medicines.

There really is bismuth in stomach-relief medicines.

Boron: Boric acid was used to treat vaginal infections for more than one hundred years. It is used as an antiviral and antifungal treatment against *Candida albicans* and the more resistant *Candida glabrata* yeast strains. Boron-based drugs show promise as therapeutic agents with anticancer, antiviral, antibacterial, antifungal, and other disease-specific treatments. A boron-based drug approved in 2003 works as a proteasome inhibitor for the treatment of multiple myeloma and non-Hodgkin's lymphoma. Several other boron-based compounds show promise in various phases of clinical trials.

Calcium: Calcium is a key component of antacids, where it buffers the pH level in the stomach and treats acid indigestion. Your body needs calcium to maintain strong bones and teeth, adding hardness. Calcium helps muscles move and enables nerves to carry messages from the brain. The strength and structure of your skeleton comes from a form of calcium phosphate called hydroxyapatite, $Ca_{10}(PO_4)_6(OH)_2$.

Copper: Copper compounds work for anti-inflammatory, anti-proliferative, biocidal, and other uses. Copper radioisotopes are suitable for nuclear imaging and radiotherapy. Nanotechnology opened new possibilities for design of copper-based drugs and medical materials. To date, copper has not found many uses in medicine, but ongoing research, as well as preclinical and clinical studies, offers hope.

Gold: Elemental gold and gold compounds have medicinal uses. Gold is an effective medicine for controlling some types of arthritis and related diseases. In some people it helps relieve joint pain and stiffness, reduce swelling, and heal bone damage, as well as reduce the chance of joint deformity and disability. Scientists developed injectable gold compounds to treat rheumatoid arthritis, but that use has faded.

Lithium: Lithium is a soft, silvery-white alkali metal effective in treating conditions such as mania and bipolar disorder. Lithium is in a class of medications called antimanic agents. As a side note, it is occasionally dissolved in certain hot springs, and in small amounts can provide a mild sense of well-being.

Magnesium: Magnesium is the fourth most common mineral in the human body after calcium, sodium, and potassium. The National Institutes of Health (NIH) says that magnesium is an essential element required for more than 300 enzymatic reactions. Without sufficient magnesium, those biochemical processes can suffer, and emerging evidence indicates that nearly two-thirds of the population in the Western world is not getting enough.

Mercury: Mercury's chemical symbol, Hg, comes from the Greek *hydrargyrum*, meaning "liquid silver." For centuries mercury was an important component of medicines and cosmetics, and two primary uses were the disinfectant mercurochrome and as a binding agent for dental amalgam. Both usages are discouraged now. The Egyptians used mercury in tombs, as did the Chinese. Mercury was thought to prolong life, so China's first emperor, Qin Shi Huang, tried ingesting it—instead of enjoying a long life, he was dead at age forty-nine. Mercury was used in very small amounts as a preservative or antibacterial agent in products such as antibiotics, blood pressure cuffs, contact lens solutions, diuretics, ear and eye drops, eye ointments, and hemorrhoid relief ointments. Its use is generally being phased out, even as its presence in skin-lightening creams has increased. (For safety's sake, avoid products containing mercury.) Even the use of mercury in thermometers is lessening due to the accuracy and safety of modern digital thermometers.

Selenium: Selenium is an element whose name comes from the Ancient Greek *selḗnē*, for "moon." It is found in soil and in certain foods such as whole grains, Brazil nuts, sunflower seeds, and seafood. Selenium is needed for proper thyroid and immune system functions. A deficiency can cause problems, but so can too much. Selenium may treat Hashimoto's thyroiditis, an autoimmune disorder, and high cholesterol. Very few uses for selenium have been approved by the US Food and Drug Administration (FDA).

Silver: Silver nitrate ($AgNO_3$) is a topical solution used in treating wounds and burns on the skin because it works well to fight infections. If you've heard of the phrase "Born with a silver spoon" it refers not just to trust-fund babies but also to the health benefits of eating from silverware. Due to its antibacterial and antimicrobial properties, actual silverware helps fight against germs. So don't lock up the good silver for special occasions—use it every day!

Sodium: Sodium chloride is widely used in nasal saline sprays, intravenous flush solutions, and eyewashes. Other forms of sodium in medicines include sodium alginate for granulation and disintegration and sodium benzoate for preservation. We'll talk more about the dangers of sodium below.

Sulfur: Sulfur is present in all living tissues, behind only calcium and phosphorous in abundance. Sulfur occurs naturally in garlic, onions, and broccoli, giving those vegetables antifungal and antibacterial properties. It has long been taken by mouth for a range of maladies, such as shortness of breath, allergies, high cholesterol, clogged arteries, respiratory infections, and more.

Zinc: Zinc was probably named by the alchemist Paracelsus for the German word *Zinke*, meaning "prong" or "tooth." It is used today in sunblock products to prevent skin cancer, but it is also an essential prenatal mineral and vital for growing children. The *British Medical Journal* estimates that some 2 billion people have some form of zinc deficiency, leading to growth retardation, delayed sexual maturation, infection susceptibility, and diarrhea. Conversely, excess zinc can lead to lethargy and ataxia.

MEDICINAL CLAY

The Ebers Papyrus is an Egyptian medical text dated to about 1550 BCE. Georg Moritz Ebers was a German Egyptologist who acquired the papyrus in 1873 and set about translating it. He listed 700 folk remedies for various ailments, such as adding malachite to small cakes to expel roundworms. Many of the treatments in the papyrus have instructions for fighting various parasites, giving us an idea of the daily travails of life under the Pharaohs.

It's true that folk remedies and ancient medicine often got results. Equally true is that some early "doctors" had more in common with priests, basically conducting exorcisms. Ancient Babylonians believed many illnesses to be related to evil spirits, and the most ancient form of surgery, trepanation, involved drilling a hole in the skull to release demons. The earliest tools for this procedure included flint knives, and presumably an opiate for a sedative. Evidence from skull analysis shows many patients throughout history survived the ordeal, according to Charles C. Gross in his article "A Hole in the Head: A History of Trepanation," published in the MIT Press Reader.

Since clubbing wounds, serious falls, and other head injuries often result in blood clotting, trephination was actually an effective treatment for the time, but Stone Age physicians also employed skull drilling for headache, epilepsy, and mental disorders.

Clay remains one of the geological marvels still in use today. "Clays are little chemical drug-stores in a packet," according to researchers at Arizona State University in 2008. Seeking a new weapon in the battle against methicillin-resistant *Staphylococcus aureus* (MRSA) infections, the team explained that clays have been used for millennia to fight infections and settle indigestion, and they noted that the Egyptian queen Cleopatra, famed for her supposed beauty, used clay facials. The practice continues today in numerous spas that feature mud baths.

The ASU scientists gathered clay samples from twenty different places around the world and investigated their strengths as an antibacterial agent. More studies are needed to identify the specific mechanism or active ingredients, but two clays from the United States and one from France showed promise.

Most modern clay treatments for facials involve baking the raw sample to remove microorganisms, but some of the therapeutic benefit has been shown to lessen with overheating. Another variable is the presence of trace minerals. One common facial treatment, montmorillonite clays, has more than seventy different trace minerals, making it hard to isolate the source of the benefits and duplicate them in a laboratory setting. Bentonite, which is rich in montmorillonite, is used externally as a wet poultice applied to skin infections or wounds, and internally as a bulk laxative. For years, over-the-counter brand medicines such as Kaopectate, Rolaids, and Maalox all contained kaolin, but manufacturers were persuaded by the FDA to discontinue its use.

QUICK CLOT

One of the scariest situations you're likely to face in a survival situation (or around the homestead) is a dripping, bloody wound. A severe hemorrhage is a frightening situation for the untrained, but there's a geology hack for that. One of the first field dressings developed to spur blood clotting was based on zeolites, ring-shaped molecules that quickly stop bleeding. Developed in 1984 and used in Afghanistan and Iraq by US soldiers, it proved to be a blessing. Unfortunately, zeolites also run quite hot when reacting with blood, and that exothermic reaction can cause second-degree burns.

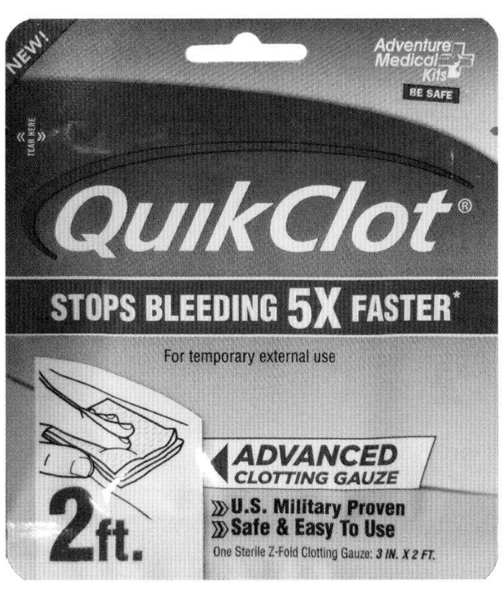

Modern quick-clot gauze contains kaolinite, which replaced zeolites.

Zeolites spur the release of calcium ions into the blood, which in turn promotes clotting. Unfortunately, zeolites have the potential to cause skin burns and are also difficult to clean from the wound. Improvements to the original formula include pre-hydrated crystals that don't react to water in the blood with the same amount of generated heat.

The solution so far has been to switch to kaolin, starting in 2012, which also spurs the body's clotting mechanism but without the heat reaction. Impregnating gauze with loads of kaolin is now done extensively in emergency rooms, and is standard issue for police cars, ambulances, and combat medics.

To find kaolinite locales near you, go to mindat.org—the mineralogical database—and run a search. It is a common mineral present in almost every US state, and occurs globally as well. ***Note:*** You can use mindat.org as a resource for any common mineral, so anyone looking for a specific mineral locale should get familiar with this free resource.

KAOLIN FOR SKIN CARE

Kaolin clay is available in bulk from websites such as KindredHomesteadSupply.com. They promote kaolin as a gentle white clay that works as a natural detoxifier. And with a high silica content, it tends to smooth the skin without sucking all the oils and moisture out. Here are some of the benefits they tout:

- **For an acne treatment,** blend kaolin, bentonite, titanium dioxide, hazelnut oil, lavender essential oil, and a small amount of tea tree essential oil. Apply to infected areas and allow to sit for 30 minutes or longer to help pull spots to the surface.
- **For scalp and hair cleansing,** mix kaolin clay, apple cider vinegar, argan oil, and tea tree essential oil and use one to three times per week as a treatment.

Even volcanic clay has health and beauty uses.

Kaolin can be added to products such as face masks, body wraps, body scrubs, soap bars, and foot treatments.

Other facial cleansers, scrubs, and masks use volcanic ash mixed with different ingredients.

INTERNAL KAOLIN USES

During a cholera epidemic at the close of the Balkan War in 1900, doctors attempted a treatment using China clay, rich in kaolinite ($H_4Al_2SiO_9$). In *Kaolin in the Treatment of Asiatic Cholera*, a report for the Proceedings of the Royal Society of Medicine in 1921, R. R. Walker writes that the kaolin treatment for cholera drove mortality from 60 percent to 3 percent. The idea wasn't new, as the Balkan doctors understood. "This salt was in use in early Roman times, and was also used by the natives of the Orinoco. It has been employed [for] diphtheria in Germany, as a mixture internally. It has been used in ptomaine poisoning with success . . . the general effect of the salt seems to point to the adsorption of toxins." The method was to dissolve equal amounts of cold water and kaolin—100 grams of kaolin dissolved in 250 cubic centimeters of water. The patient then drank a half pint of the solution every 30 minutes for 12 hours. "Vomiting soon ceased, the pulse improved, and the patient slept," concluded the report.

Caution: You should know exactly what you're doing before trying any of these medicines on your own. This book does not contain medical advice, just interesting information about practical geology!

GEOPHAGIA

Geophagia, also known as geophagy, comes from the Greek words for "earth eating." It is the intentional practice of eating earth or soil-like substances such as clay, chalk, or even termite mounds. The behavior can range from harmless experimentation to a full-grown obsessive compulsion. The NIH quotes the American Psychiatric Association in this description:

"From a psychiatric point of view, geophagia has been classed as a form of pica—a term that comes from the Latin for magpie, a bird with indiscriminate eating habits. In its *Diagnostic and Statistical Manual*, the American Psychiatric Association defines pica as persistent eating of non-nutritive substances that is inappropriate to developmental level, occurs outside culturally sanctioned practice and, if observed during the course of another mental disorder, is sufficiently severe to warrant independent attention."

More than one hundred primate species have been documented to supplement their diets with mineral-rich soil. Tribal cultures and traditional rural societies around the world have certain rituals and

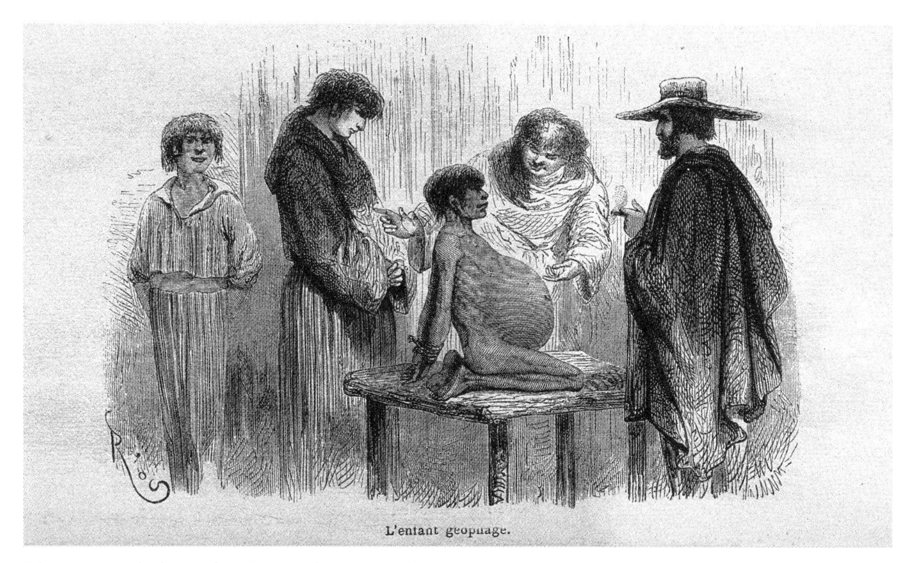

Friars at Pevas (Pebas, Pehuas), Peru, showing Paula Marcoy, a 5-year-old girl with a desire to eat earth (geophagia). Wood engraving by E. Riou WELLCOME LIBRARY, LONDON, REFERENCE NO. 571257I, 1860 (IN THE PUBLIC DOMAIN)

practices that are either harmless or mildly beneficial. Interestingly, the practice has never been documented in Japan or Korea. Historically, Pliny documented the ingestion of soil on Lemnos, an island of Greece, and Hippocrates (460–377 BCE) mentions geophagia. Alexander von Humboldt recorded that the Otomac of South America used soil regularly in their diets.

In a 2012 essay titled "Geophagy: An Anthropological Perspective" for the book *Soils and Human Health*, authors J. M. Henry and F. D. Cring explain that clay minerals have been reported to have beneficial microbiological effects, "such as protecting the stomach against toxins, parasites, and pathogens." It turns out that we humans are unable to synthesize vitamin B12, so geophagia may be a behavioral adaption to obtain it from bacteria in the soil. Mineral content in soils may vary by region, but many contain high levels of calcium, copper, magnesium, iron, and zinc—minerals that are critical for developing fetuses and that can cause metallic, soil, or chewing ice cravings in pregnant women.

Geophagia should not be confused with the practice of rock-licking. When collecting pebbles in the field, many "Golden Age" rockhounds of the 1960s and 1970s would lick a rock to see what it might look like when polished. The wet surface shines the rock and reveals how it might appear when the patina or dust is gone. The problem is, there many germs and diseases out in the wild that you can pick up, including hantavirus, giardia, amoebic dysentery, and any number of other complaints. The modern solution is to bring a spray bottle.

Patients exhibiting extreme geophagia can often be reluctant to admit their problem until confronted with X-rays showing intestinal blockages. Other issues stem from parasites living in soil and unintended mineral imbalances. At the farthest end of the spectrum, sufferers can display peritonitis and tears in the intestines.

THE IMPORTANCE OF SALT

Dating to the earliest times, salt (NaCl) has been an important part of human life. It makes roasted meat taste better, it preserves a bountiful harvest for future use, it sparks up soup and stew, and it is important to the human body in the right doses. As early as 6050 BCE, salt was known to historians—the Egyptians used salt in religious ceremonies and traded it with their far-flung Mediterranean outposts.

Language specialists point out that the word "salary" derives from the Latin *salarium*, which also means "salary" and has the root *sal*, or "salt." In ancient Rome, *salarium* specifically meant the amount of money allotted to a Roman soldier to buy salt, which the army often used when gold or other currencies were in short supply. The term "worth their salt" implies that the worker performed their task suitably, equal to the amount of salt paid out. The Romans even sprinkled salt on their leafy vegetables—hence the word "salad."

Monopolies often controlled the production of salt, either by evaporating seawater, boiling the water from salty springs, or mining sedimentary salt deposits. Production was legally restricted in ancient times, and salt was historically used as a source of trade and currency.

In his epic book *Salt: A World History* (Penguin Books, 2002), Mark Kurlansky sets out a fantastic review of the importance of salt to medicine, cooking, food preservation, economics, and more. It's a fantastic read—the late Anthony Bourdain proclaimed the book a "must-have for any serious cook."

Kurlansky lists fascinating detail after detail. Early Chinese governments maintained a monopoly on salt prices by strictly controlling production. Most Italian cities sprang up around saltworks. The first of the great Roman roads was the Via Salaria, or Salt Road. During the Punic Wars with rival Carthage (264–146 BCE), Rome manipulated salt prices to raise money for the long campaign, and the official of the treasury became known as the *salinator*.

Rome developed or conquered saltworks wherever it went. "By conquest they took over not only Hallstatt, Hallein, and the many Celtic works of Gaul and Britani but also the saltworks of the Phoenicians and the Carthaginians in North Africa, Sicily, Span, and Portugal," Kurlansky writes (63). In all, he reports, experts identified more than sixty ancient saltworks from the Roman Empire.

Ancient uses for salt include preserving most meats, particularly for curing ham and pork products; preserving fish such as sardines, tuna, mackerel, and bonito; and for safely storing olives, fruits, and vegetables. They developed a fishy, salty sauce called *garum*, which physicians prescribed for digestive disorders, sores, sciatica, tuberculosis, and migraine headaches. A rough Chinese equivalent would be soy sauce, fermented over great lengths of time and used liberally to add a salty flavor to many dishes.

The Romans also learned how to create the unique purple dye of royalty when they discovered the link between salt, a Mediterranean mollusk, and the Phoenician salt fish trade. "From as early as 1500 BCE, this dye brought wealth to merchants in Tyre," Kurlansky writes (Kurlansky, 76).

There are so many anecdotes and fun facts in Kurlansky's book that they almost tumble forth. In colonial America, salt springs and licks often paved the way for settlements, expanding on trails created by animals in search of salt. Buffalo, New York, was founded at the end of a wide trail trampled out by buffalo seeking salt. Bering Straits Eskimos boiled reindeer, walrus, bear, sheep, and other game in seawater to add salt to the meat. The Delaware Indians salted their cornmeal, and the Hopi boiled beans and squash in salted water. "As on the Italian peninsula, all the great centers of civilization on the American continents were founded in places with access to salt," he writes (Kurlansky, 203). The Inca exploited salt wells near Cuzco, and the Chibcha salt lords in Bolivia swore off sex and salt twice a year to honor their gods.

Mayan saltworks date to at least 1000 BCE. "The Mayans used salt as medicine mixed with marjoram and xul tree leaves for birth control, with oil for epilepsy, [and] with honey to lessen childbirth pains," Kurlansky continues. He notes that Mayans far removed from the ocean also learned how to extract salt from the leaves of certain palm trees by burning the leaves and soaking the ashes.

Homegrown salt crystals

In his "American Salt Wars" chapter, he describes how English and Spanish seafarers would slip ashore at certain Caribbean islands to rake salt, fill their ships, and dash off. At times the trade in salt back to Europe outweighed shipments of sugar, rum, or molasses. Under the heading "The Mythology of Geology," he describes the birth of canning and preserving food, including salt fish and vegetables, then pivots to the advantages of salt domes to trap petroleum deposits.

The downside of so much salt in our diets eventually became clear, and another noted historian, Jared Diamond, explains it well in his book *The World Until Yesterday: What We Can Learn from Traditional Societies* (Penguin Books, 2012). In the chapter titled "Salt, Sugar, Fat and Sloth," he explains that "most plants contain very little sodium, yet animals require sodium at high concentrations in all their extracellular fluids. As a result, while carnivores readily obtain their needed sodium by eating herbivores full of extracellular sodium, herbivores themselves face problems . . . that's why the animals you see coming to salt licks are deer and antelope, not lions and tigers" (Diamond, 415).

Diamond contributes multiple pages to the link between sodium intake and health issues. He cites 120 over 80 as the average blood pressure reading for American citizens, then shows that Brazil's Yanomamo Indians average a remarkable 96 over 61. They also have the world's lowest average daily salt intake, about 50 mg, and strokes are exceedingly rare among this tribe. Conversely, Japan is dubbed "The Land of Apoplexy" in medical circles because strokes are the leading cause of death by a factor of five over the United States. In the village of Akita in northern Japan, salt consumption can range from an unhealthy 5 grams per day to a staggering, heart-stopping 61 grams per day. The average blood pressure measured at Akita is 151 over 93, and few villagers live beyond seventy years of age (Diamond, 419).

You'll need a solid hammer to break off pieces of this hardened mudstone.

CHAPTER 10

IMPROVING THE HOMESTEAD

With the basic survival challenges overcome, the next step is to set up some kind of permanent base camp. We'll jump from long-term camp to homestead in the same breath, as the concepts are basically the same.

PICKING A PLACE

In *The Survival Book* (Funk & Wagnalls, 1968), Nesbitt, Pond, and Allen offer excellent advice. "Pick your location carefully . . . try to be near fuel and water. Water supply is probably the most important factor in selecting a camp site. Pick a location protected from the wind but away from the dangers of rock falls, avalanches, or floods. In other words, use common sense" (Nesbitt et al., 31).

In 1805 adventurers Meriwether Lewis and William Clark faced the challenge of building a solid fort to house their soldiers through the bleak winter at the Oregon coast. In *The Journals of Lewis and Clark* (Houghton Mifflin, 1953), editor Bernard DeVoto wrote that the captains wanted to avoid the noise of the ocean waves, the spray of salt water,

This replica of Fort Clatsop provides a good idea of how to construct an off-the-grid compound.

and set some distance between them and local villages. Clark selected a site about 7 miles up a creek, "in a thick groth of pine . . . on a rise about 30 feet higher than the high tides leavel" (DeVoto, 294).

In his classic book *How to Survive Off the Grid* (WeldonOwen, 2019), Tim Macwelch suggests strongly considering access to water, windbreaks, and a southern exposure. "Most of today's home builders aren't in touch with the seasons and the environment [the way] builders were a few centuries ago," he writes. "Take the time to consider the importance of the sun's position and prevailing winds. Early Americans . . . knew that building on an east–west axis provided great advantages" (Macwelch, 22). The benefits include passive solar heat through south-facing windows as well as the ability to open windows and clear out dampness or odors.

WHERE TO LIVE SAFELY

Thanks to multiple government agencies involved in land-use planning, it's fairly easy to locate a geology map for your area. Many states have excellent departments of geology, mining, or related agencies; the US Geological Survey is another solid resource.

What you want to be careful about are floodplains, earthquakes, and radon. In many jurisdictions you can find hazard maps for these threats.

Radon

Radon is a radioactive gas that kills an estimated 14,000 US citizens from lung cancer every year, according to the US Environmental Protection Agency. It forms from the breakdown of naturally radioactive metals such as uranium, thorium, or radium found in granite, basalt, and other rocks. Radon-222 is the decay product of radium-226, and they are both part of the long decay chain for uranium-238. Since uranium is present in varying amounts just about everywhere in the Earth's crust, radium-226 and radon-222 are also present in almost all rocks and soils.

The half-life of radon is only 3.8 days. The amount of radon in the soil depends on soil chemistry, which varies from one house to the next. Radon levels in the soil range from a few hundred to several thousand pCi/L (picocuries per liter) in air. The amount of radon that escapes from the soil to enter a building depends on the weather, soil porosity, soil moisture, and the suction within the building.

People can be exposed to radon primarily from breathing it in air that comes through cracks and gaps in buildings and homes.

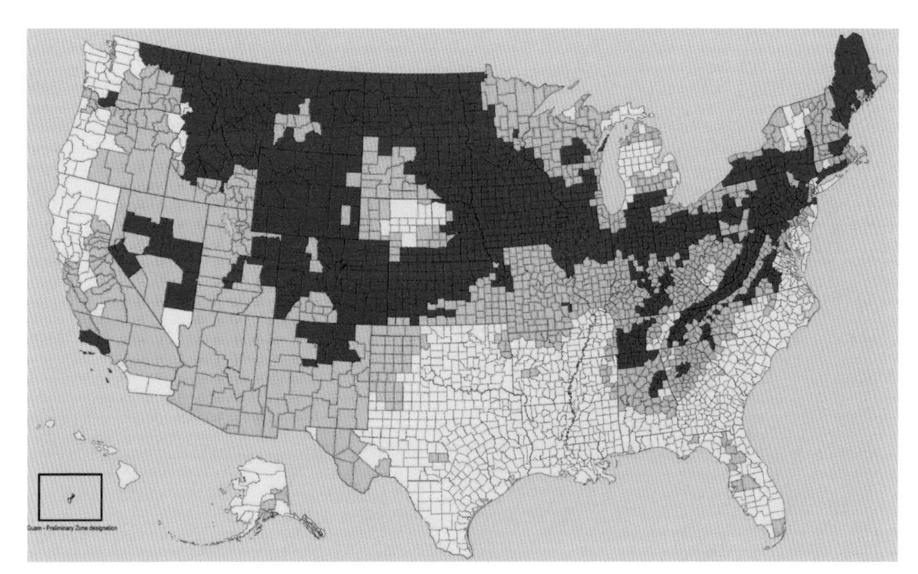

Radon map for US counties. Zone 1 counties, in red, show greater than 4 pCi/L; Zone 2, colored tan, 2–4 pCi/L; Zone 3, yellow, is safest, at less than 2 pCi/L. Modified from EPA.gov.

The Map of Radon Zones was developed in 1993 to identify areas of the United States with the potential for elevated indoor radon levels. The map is intended to help governments and other organizations target risk-reduction activities and resources. The Map of Radon Zones should not be used to determine if individual homes need to be tested. No matter where you live, test your home for radon—it's easy and inexpensive. There is danger if your radon level is 4 pCi/L or higher. Consider fixing if your level is between 2 and 4 pCi/L.

Earthquakes

The potential for earthquake damage varies across states but is especially high in two regions—the West Coast and the Midwest. The southern Alaska coastal region also contains high-risk zones. If you can't avoid these areas, you should at least consider adding additional earthquake-resistant features when building structures.

Flooding

The best soil for farming historically drew settlers and farmers to fertile valleys, but the trade-off was greater risk of damage due to flooding. The Federal Emergency Management Agency (FEMA.gov) has an impressive number of free maps available for checking on the specifics of your area, plus an interactive map to zoom down to your location.

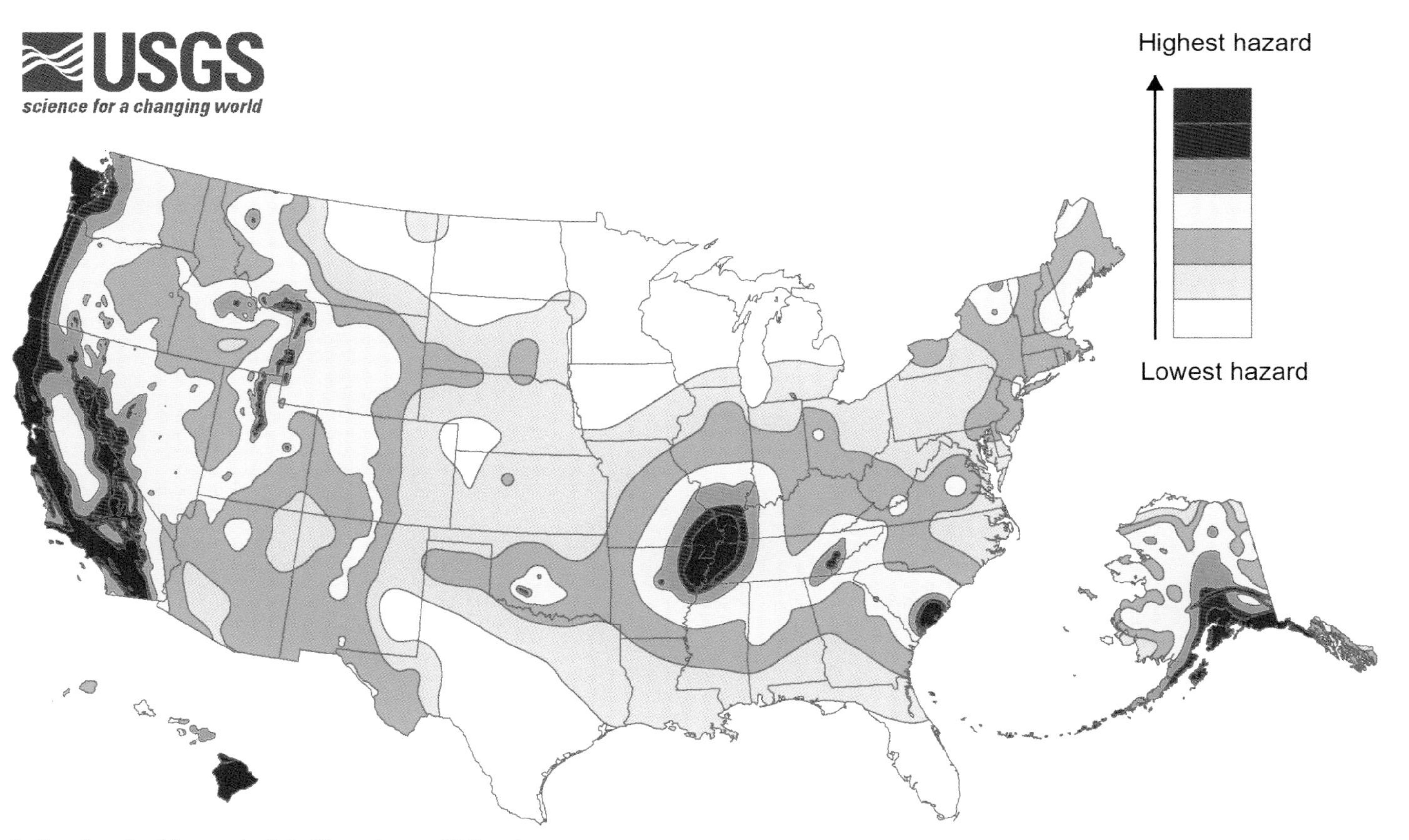

Earthquake potential across the United States (source: USGS.gov)

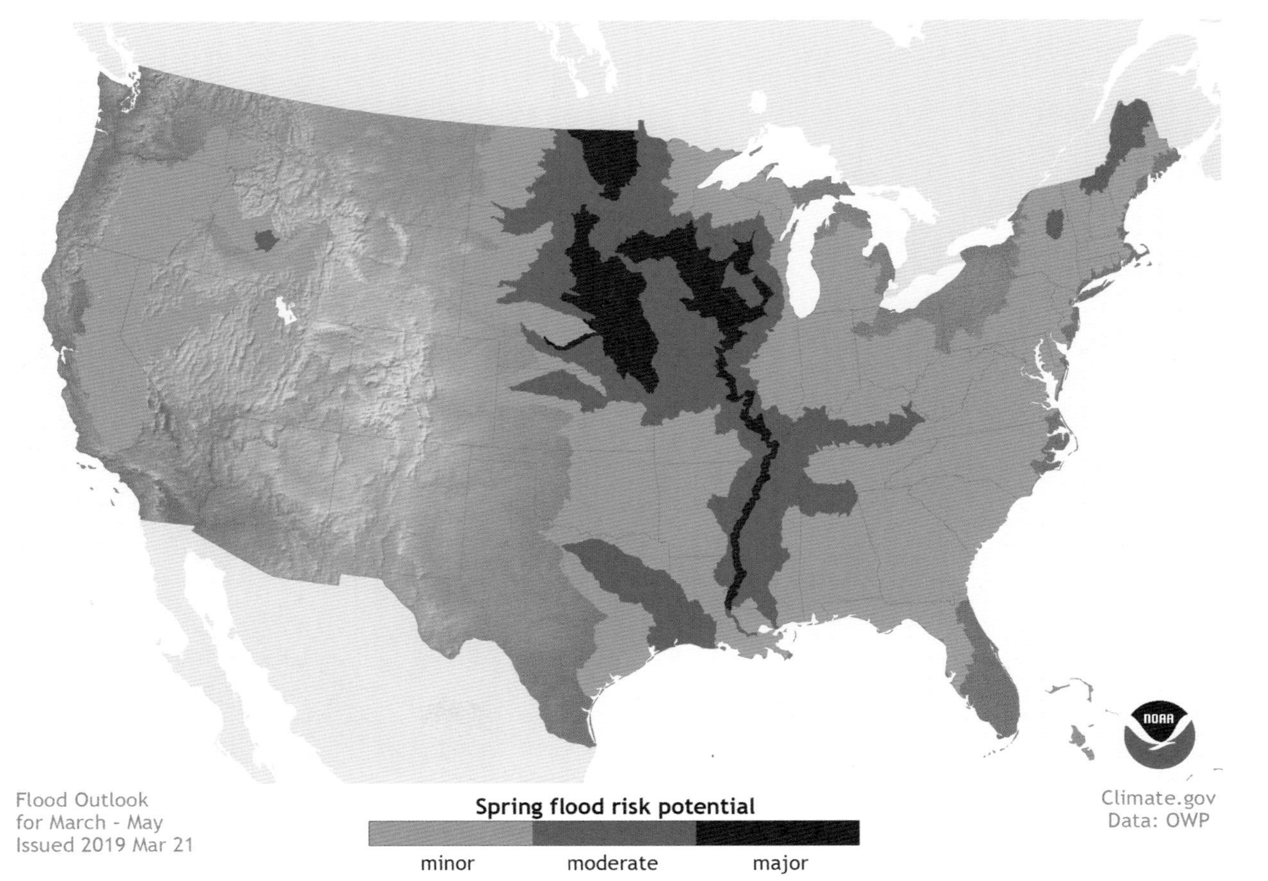

Example of spring flood risk potential from 2019 (source: Climate.gov)

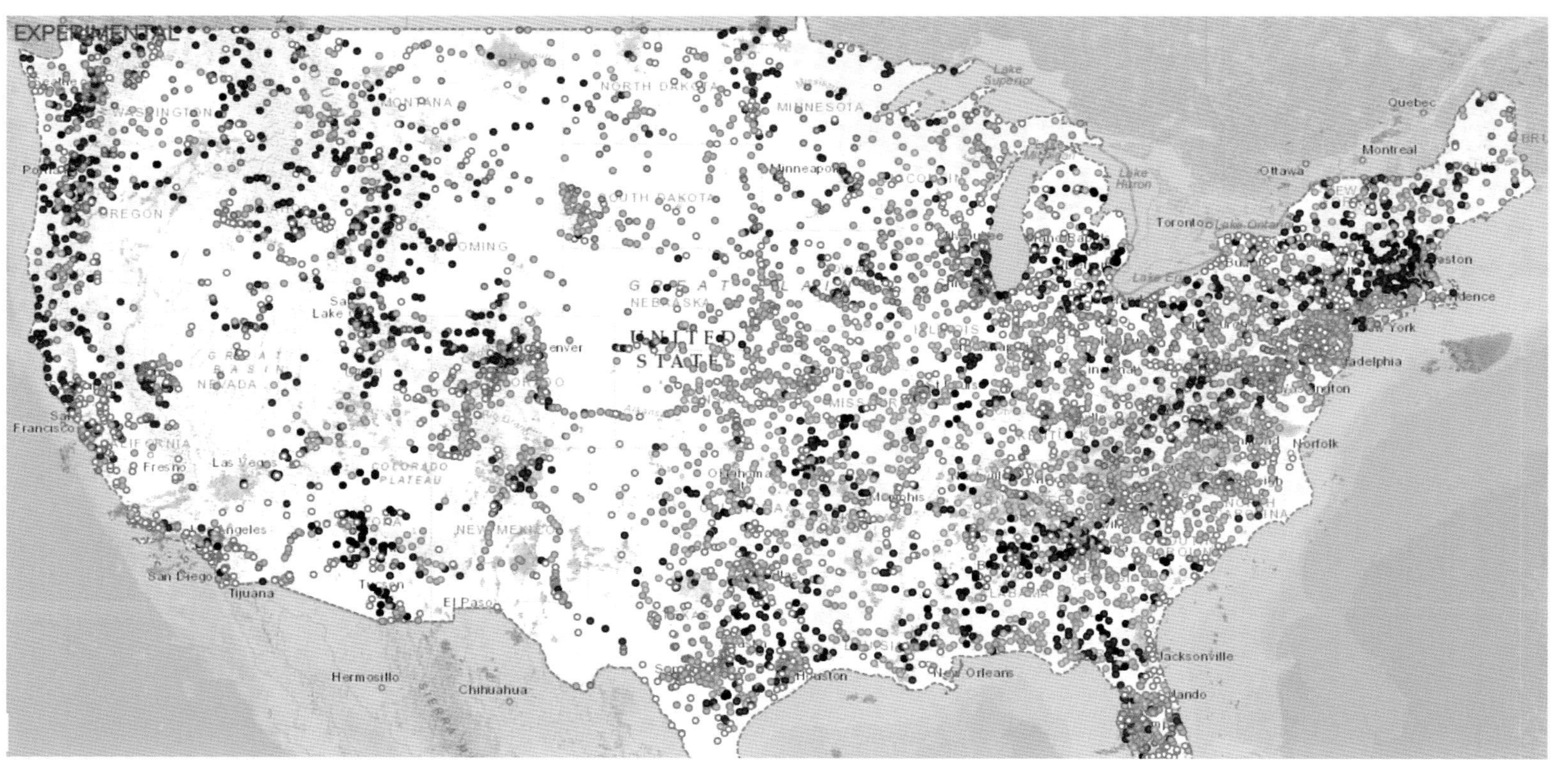

Experimental interactive map of USGS water level gauges (source: USGS.gov). Dark blue dots depict higher than average water flows; red indicates the most extreme drought conditions.

Your state and local government agencies have additional information. Climate.gov provides regular updates for spring flooding risks, such as the 2019 map on page 166.

You can also access specific water level gauges at WaterData.usgs .gov, providing access to specific gauges.

Volcanoes

Living near an active volcano is something every homesteader should avoid. While volcanic ash converts to excellent soil, the dangers range from glowing ash clouds, inundation from blowing ash, earthquakes, flooding, and, in Hawaii, even glowing lava.

Most of the volcanic activity in the United States is in two areas—the Kilauea Volcano in Hawaii and the West Coast, particularly the active peaks in the Cascade Range. You can find recent lava flows, about 1,000 years old, along the Pacific Crest Trail, for example, in northern California, Oregon, and Washington.

In Hawaii, Kilauea's last major eruptive cycle ended in 2018, but by late 2021 the volcano was showing renewed signs of life. That 2018 lower Puna eruption caused twenty-four injuries, forced extensive evacuations, and caused about $800 million in property damage.

One of the biggest dangers facing the United States comes from the impending eruption of the Yellowstone Caldera, which is North America's most potentially dangerous super volcano. Yellowstone is not overdue for an eruption, but the question isn't really IF, but WHEN. Yellowstone has experienced three major eruptions: 2.08, 1.3, and 0.631 million years ago. This comes out to an average of about 725,000 years between eruptions, but with a delta of hundreds of thousands of years. Based on math, you could predict we'll be right on schedule in 94,000 years.

While interim "minor" hiccups are much more likely, there is great risk of an eventual "super-eruption" capable of affecting weather on a global scale. The 1980 eruption of Mount St. Helens released about 0.25 cubic kilometer of material called tephra. Yellowstone's Huckleberry Ridge eruption about 2.1 million years ago released an estimated 2,450 km^3; the Toba eruption 74,000 years ago may have been even larger. That would be enough to bury Wyoming, Montana, Idaho, and Colorado under 3 feet of ash.

Plinian eruption column from Mount St. Helens, May 18, 1980 (aerial view from the southwest)
PHOTO BY ROBERT KIMMEL, COURTESY OF US GEOLOGICAL SURVEY (IN THE PUBLIC DOMAIN)

Old Faithful eruptions at Yellowstone National Park are a crowd-pleaser, but they indicate the potential for widespread ash fall damage.

Mantle Creep

Mantle creep, also called soil creep, is a threat to any hillside homestead. In general, it is the slow, downward movement of soil over time. Creep can be extremely slow but frustratingly persistent, and almost impossible to correct. Basically, you have tons of earth obeying the law of gravity. It's like a slow-motion landslide.

Even moderate slopes of 10 percent can show signs of creep. Look for irregularities in drainage, soil distribution, vegetation cover, and, in particular, trees bending at their stump. As the soil continues to push downward, it forces the tree to arc back uphill. Because the tree can react and continue its regular upward push, a large bend develops pointing downhill, opposite the direction of the creep. This is the easiest diagnostic you'll find. Other telltale signs are growing cracks in foundations, ruptured in-ground pipes, and uneven settling in the foundation.

Bent trees at the base display telltale signs of mantle creep.

FIREPLACES AND HEARTHS

Building a long-term shelter requires the ability to cook and heat indoors. You may need distinct areas to keep those chores separate over long periods of time, but in the short run, a simple fireplace may serve both needs.

For example, you can construct the fireplace at the same time you erect the shelter, piling up rocks with a lip to keep coals safe on the floor, piecing in rocks as needed, and positioning logs with fire safety in mind.

For safety, you want to guard against hot coals building up in cracks and crannies, where they might eventually come into contact with wood. You can place wet mud or damp soil between rocks, around wood ends, and as needed to avoid drafts.

A good example of controlling airflow around a central fire pit comes from the cliff dwelling ruins at Mesa Verde in southwestern Colorado. A series of clay panels about waist-high or less may have been used to reflect heat, perhaps augmented by stretched hides or other movable structures, which could also adjust airflow.

If you dig into the soil in a dugout or pit, the soil bank offers good options for dispersing smoke. Keep in mind that over time, hot smoke will dry out soil and wood, as well as roots and foliage. The key safety concern is setting the entire structure on fire, so situate your wood supply away from the flames, keep the overall fire level low, and constantly check for hidden smoldering embers.

In larger structures, your fireplace or hearth will require considerable thought and preparation. Mud is a good short-term solution for setting stones, but you'll need concrete or mortar for more permanence. The same consideration applies to using bricks—you'll need a good mortar to hold the structure in place and weather the extreme temperature swings you are sure to encounter.

Laying stone fireplaces and chimneys requires skill, patience, and experience. You can find plenty of resources online or in books to guide you. There are several interesting YouTube videos as well. The payoff is clear, however: Stone structures last far longer, and usually far outlive the wooden structures built around them.

Ancient traditions abound for creating a hearth. At SongOfAmergin .wordpress.com, author Brian Walsh recounts how the Celtic hearth was the heart of the household. "The focus . . . was the household hearth," Walsh wrote. "It was not only where food was prepared, but also the primary source of warmth and light for the dwelling, around which people gathered through the long winter nights."

Walsh explains that the hearth was traditionally in the middle of Iron Age round houses. It rested under the vertical ridgepole and vented through the capped smoke hole. Throughout the day the fire burned or smoldered as needed, and banking it to last through the night was a key chore, often accompanied by prayers for protection.

According to Walsh, should the fire go out, that was a bad sign of ill luck from the hearth goddess. For Catholics, the first fire in a new

Inside a crude cabin with circular fire pit and ruins of stovepipe for smoke control

Fire pit and heat control baffles in an excavated kiva at Mesa Verde cliff dwellings

hearth was a very big deal. Summoning protection for a new hearth was an important ritual, sometimes accompanied by burying a piece of holy iron.

Immense lodge hearth using rough, angular stone serves as a central feature to the main gathering area.

The hearth or fireplace was often the dramatic centerpiece of the building, and it dominated the main gathering area for lodges and castles throughout history.

There are countless variations for stunning fireplaces, so the design you settle on will depend on many factors—expense, resources at hand, experience, access to expert crafters, etc. Do your homework. Consider experimenting with mortar, showing off the rocks.

Massive lodge fireplace with rounded river stones set with minimal mortar showing

OVENS

Fireplaces and stone circles have their place, but humans have always had a knack for continual improvement. The construction of a dedicated oven would have been logical for better temperature control, useful for slow cooking, roasting, and baking.

Recent research from an excavation team led by Jiˇr´ı Svoboda described an Upper Paleolithic roasting pit at the center of an area about 5 meters (16 feet) across. Published in 2008 in the journal *Antiquity*, "Pavlov Vi: An Upper Palaeolithic Living Unit" describes the site found in the Czech Republic. Thanks to clear stratigraphy, radiocarbon dates, faunal and floral remains, transported rocks and tertiary shells, stone and bone artifacts, decorative items and ochre, the researchers were able to pinpoint the date of occupancy to about 29,500 years ago.

The team quickly realized that the pit's gigantic dimensions served an obvious purpose—to roast giant portions of mammoth meat. The majority of the identified bones belonged to woolly mammoths (*Mammuthus primigenius*), primarily represented by two individuals: an adult female between twenty and twenty-five years of age and most likely her calf, between two and five years of age.

According to the report, the settlement had a varied diet. "The site also contained the remains of horse, reindeer, wolf, fox, wolverine, bear and hare," and included a "lithic industry" of artifacts, decorative objects, and pellets of burnt clay. The team found ceramic pieces and fragments that bore witness to human activity with a variety of fingerprints. There were magical or ritual fetishes, a carnivore head, and tools for textile industry. The tools consisted of flint-pointed blades, burins, and end scrapers. The builders of the pit used cobblestones from nearby, clearly brought intentionally for construction.

Ancient Egyptians, Romans, and Asians all used stone or brick designs. Many resemble a modern pizza oven, capable of temperatures to 500°C (932°F).

At the Universal Appliance and Kitchen Center website (uakc.com), historians place the credit for the first official oven as a brick contraption built in 1490 in Alsace, France, using brick and tile. Around 1728, German-designed cast-iron ovens, known as five-plate or jamb stoves, made their appearance. Next, a Bavarian architect named François de Cuvilliés designed the first recorded enclosed oven, which served to control smoke. It was called the Castrol stove but was known as a stew stove. Finally, about seventy years after the jamb stove made its appearance, kitchens welcomed the creation of an iron stove with temperature controls for the stovetop.

KILNS

The technology of controlling fire evolved from furnace to kiln thousands of years ago. The word derives from the Old English *cylene* (ky-LEEN), which was borrowed from the Old Welsh *cylyn*, which, in turn, was borrowed from the Latin *culīna*, which means "kitchen," "cooking stove," or "burning place."

Kilns started as a simple earthen trench filled with pots and fuel, a process known as pit-firing. Pit-firing is the oldest known method for the production of pottery, with raw material set in a trench, covered, then allowed to heat under a large fire for considerable time—sometimes days. One of the first improvements was to construct a central firing chamber with baffles and a stoking hole, which saved on fuel. Another refinement was to build a chimney stack so the kiln could draw air easily, allowing higher temperatures and better fuel combustion.

According to P. M. Rice's article "On the Origins of Pottery" in the *Journal of Archaeological Method and Theory*, experts dated pit-fired pottery to 29,000–25,000 BCE. The earliest known kiln dates to around 6000 BCE and was found at the Yarim Tepe site in modern Iraq. The Iraq site yields important evidence of practical geology. According to *Early Stages in the Evolution of Mesopotamian Civilization: Soviet Excavations in Northern Iraq*, edited by N. Yoffee and J. J. Clark (University of Arizona Press, 1972), Yarim Tepe was apparently a large settlement, with a limestone granary, mud bins and silos, and grinding stones made of basalt, sandstone, or limestone. Projectile points made from obsidian are common there.

Chinese kiln technology was the most advanced in the world for centuries. The Chinese developed kilns capable of firing at around 1,000°C (1,832°F) before 2000 BCE. These were updraft kilns, often built belowground. Two main types of kiln were capable of reaching the temperature required to fire porcelain—the dragon kiln, which ran up a slope, and the smaller, horseshoe-shaped mantou kiln. Both could reach 1,300°C (2,372°F) or more.

Kilns have several advantages over fireplaces. Kilns allow higher temperatures, and also use fuel more efficiently. Some of the advantages of kilns include the following:

- Melting silica for glass
- Firing clay for ceramics
- Curing bricks

- Melting and casting metal ores
- Treating limestone for quicklime
- Reducing wood to charcoal
- Drying tobacco, barley, hops, corn, and lumber

The ruins of charcoal kilns and limestone kilns dot the western United States. These were usually industrial-sized kilns, often in a row of five or six units. Charcoal was a key resource for early ore extraction processes, and thus valuable. Compared to digging coal out of the ground, however, charcoal kilns were labor-intensive and time-consuming. In areas of the world where high-grade anthracite coal was readily available, ancient engineers would have been happy to bypass the process of reducing firewood to charcoal.

Limestone kilns supplied quicklime for construction projects. This would have been handy for any large project requiring a level floor, such as a mill for processing metal ores.

Airflow is a key concern when maintaining a kiln. Bellows forced air over hot coals and greatly increased the temperature. Kilns have tremendous use on a homestead above and beyond cooking—drying and preserving grains such as barley for later use as an ingredient for beer, for example.

Charcoal kiln from the Dolores River in the Colorado desert

Kilns come in multiple configurations, but there are a couple of major distinctions. Continuous kilns, also called tunnel kilns, process raw material as it moves through, starting and ending at room temperature and reaching high heat in the middle. Intermittent kilns work by placing raw material in the kiln, firing it up, allowing it to cool, and then removing the finished product.

From your own experience with campfires, you probably know what happens after you're sitting around burning wood for a while. Someone asks, "Does this burn?" You can easily figure out how our ancestors learned about coal, oil shale, petroleum, and other combustibles.

Pottery

At its simplest form, pottery is converting clay-rich mud into usable objects. Baking or firing the molded form at high temperature leads to a chemical reaction that hardens the form.

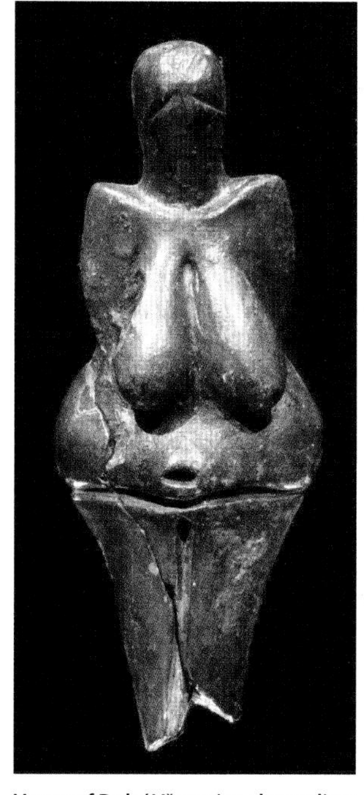

Venus of Dolní Věstonice, the earliest discovered use of ceramics (29,000–25,000 BCE) PETR NOVÁK. THIS FILE IS LICENSED UNDER THE CREATIVE COMMONS ATTRIBUTION-SHARE ALIKE 3.0 UNPORTED LICENSE, EN.WIKIPEDIA.ORG/WIKI/EN:CREATIVE_COMMONS.

The oldest known examples of pottery are figurines dating back to 29,000–25,000 BCE. Researchers excavated a ceramic statuette known as the Venus of Dolní Věstonice in 1925 in a layer of ash in Czechoslovakia. A team led by Pamela Vandiver investigated the origins and technology of the figurine for an article in the journal *Science*. They determined that the source material was a local soil known as loess, and calculated that the ceramic reached 500°C–800°C (932°F–1,472°F) in a kiln.

There are three different groups of clay-based pottery, although some researchers simply divide the technology into temperature ranges known as low-fired and high-fired.

Earthenware pottery is the crudest application, originally using open bonfires or pit fires to create plain vessels. Just about any clay will work, and the firing temperature can be as low as 600°C (1,112°F). Once artisans developed a ceramic glaze, earthenware made storing and transporting liquids practical.

Stoneware requires a kiln, reaching temperatures as high as 1,200°C (2,192°F). While Chinese artisans from early on had access to higher-quality clay and efficient kilns, in Europe the advance took longer. Chinese stoneware dates to the Tang dynasty, which ended in 906 CE. European artisans developed stoneware much later, in the late Middle Ages.

Porcelain is the highest form of pottery, requiring temperatures as high as 1,400°C (2,552°F) and uses specialized clays rich in the mineral kaolinite, a phyllosilicate with the formula $Al_2Si_2O_5(OH)_4$. The source for kaolin-rich clays is feldspar-rich rocks. The name derives from Gaoling, a village in Jiangxi province. In 1712 a Jesuit priest named François Xavier d'Entrecolles was working in China and wrote lengthy letters to his superiors in France on the advances he noted in his travels. He detailed technologies including porcelain production, raising silkworms, creating synthetic pearls, and oral vaccination for smallpox.

Terra-cotta

The term "terra-cotta" derives from the Italian words for "baked earth" and from the Latin *terra cocta*. It is a version of clay-based earthenware pottery, usually porous, and commonly used for sculpture or vessels such as flowerpots, or as water pipes and roofing tiles.

Archaeologists unearthed terra-cotta female figurines from a site in Pakistan dated as old as 3000 BCE. A terra-cotta plaque from ancient Mesopotamia dates to about 1950 BCE. Ancient Greeks used terra-cotta, as did Egyptians and the civilizations of the Indus Valley. On the other side of the world, most Olmec figurines were made from terra-cotta.

One of the most famous examples of terra-cotta art was the Terracotta Army created for the tomb of the first emperor of China in 248 BCE. Artisans mass-produced about 6,000 figures, including warriors and horses, using ten or more faces, which they augmented with ground precious stones, intensely fired white bones, dark red pigments of iron oxide, red cinnabar, green malachite, blue azurite, black charcoal, and a purple cinnabar-barium-copper-silicate mix.

Army of terra-cotta warriors discovered in 1974 in Lintong District, outside Xi'an, Shaanxi, China

MAKE YOUR OWN CLAY

Collecting high-quality clay suitable for creating pottery is a fairly simple process. Off-grid homesteader Ashley Adamant describes the process at PracticalSelfReliance.com, noting that just about any soil will do—even a sandy loam can contain up to 20 percent clay.

To get started, she advises looking for an area where rainwater doesn't appear to drain very well. Then follow this process to test your soil:

1. Scoop up enough soil to half fill a 1-quart jar.
2. Mark the height of the soil on the outside of the jar.
3. Add water to the top.
4. Stir vigorously to completely dissolve the soil.
5. Allow to settle.

After several minutes, you'll see the process of sedimentation in action. The heaviest rocks and sand will quickly settle to the bottom of the jar. A fine silt will pile up in the next layer. Anything still in suspension at this point is clay. You can estimate the percentages based on your original mark.

If you need to collect clay in dry, desert areas, you can use a dry harvest method involving collecting, grinding, sifting, and winnowing. It's very time-consuming and takes a lot of work, but if you don't have much water, that's the way to go. Strive for the smallest possible particles.

In wet areas, collecting clay is extremely simple. You can experiment with different soils around you, looking for areas that hold the most promise.

1. Fill a bucket about one-third full with clay-rich soil.
2. Add water to fill to about three-quarters or more.
3. Stir vigorously.
4. Allow to settle for a bit. After 10 minutes at the most, the rocks, sand, and silt will all separate.
5. Place a rag or old sheet across the top of a second, empty bucket. You may have to double-up the crude filter.
6. Pour the liquid so that it slowly filters through the cloth into the bucket. The wet clay remains on the cloth. You can scrape it off into another bucket as you go, or you can tie off a sheet and hang it from a tree for several hours.

The process isn't exact. You may still want to further dry out your raw clay in the sun or repeat the straining and settling process to remove more impurities. The general rule is that the longer a clay particle remains suspended in water, the higher the quality. Once you start playing with the clay, rolling it into a ball or into long cylinders, you'll learn more about what you have. It takes practice, experimentation, and patience to get it right.

Glendale University's website (Glendale.edu) has a good description of clay geology; they advise choosing clay deposits far from their source, as primary or residual clays often do not shape easily. Secondary clays, far from the source, are much more workable due to a feature known as plasticity—a similarity with plastic. They also have advice on preventing the almost inevitable cracking during firing:

"Frequently coarser clay bodies contain a particulate additive called grog, which gives the body roughness. Porcelain clays have little or no grog. Stoneware clays usually have some. Earthenware clays may or may not have grog, so this difference alone does not help us distinguish low- from high-temperature clays. Grog is commonly either sand or fired clay which has been crushed and sized. Lacking the microscopic size and shape of clay particles, grog decreases the plasticity of the clay body, but it does have a beneficial effect on shrinkage. Since it is not clay, grog does not shrink as clay does. Therefore, its presence in clay reduces the overall shrinkage rate of the clay; more grog = less shrinkage; less grog = more shrinkage."

There are numerous videos on YouTube featuring primitive technology practitioners who collect clay from termite mounds or harvest porcelain clay from near water courses. John Plant's book *Primitive Technology: A Survivalist's Guide to Building Tools, Shelters and More* (Clarkson Potter, 2019) is a must-read, and his YouTube channel is extremely popular.

The Primitive Skills YouTube channel also displays excellent knowledge, and while not as popular as John Plant, it has almost 1 million subscribers. The channel follows the exploits of a young man in Vietnam who uses geological resources to the fullest to create rice paddies, fish pens, terraced gardens, brick structures, porcelain cooking implements, crude iron forging, and much more.

MAKE YOUR OWN ADOBE BRICKS

"Adobe" is a Spanish word derived from the older Arabic *al-tob*, dating from the Moorish occupation of Spain. Mud brick, or adobe, is one of the oldest geology-based building materials. Adobe brick building is an ancient technique common in the Americas and the Middle East. The oldest structures, dating back to at least 8300 BCE, are in adobe, as are some buildings around 900 years old, which are still in use.

Adobe bricks contain a mixture of clay, coarse sand, fine sand, silt, and water. The amount of clay should be less than 20 percent, or severe cracking could result. If clay is limited, the ancients used a binder such as straw. Builders soon learned to begin the lowest course on a solid foundation such as stone, about 1 foot tall, to raise the lowest bricks away from ground moisture.

For mortar, early builders used a similar clay-sand-water mix between the adobe bricks. They soon learned to maintain large roof overhangs to protect from rain and intense sun. To protect the adobe mud mix further, builders applied two coats of lime plaster to the exterior walls.

The dwellings and walls of Jericho in modern-day Israel were constructed from mud bricks around 9000 BCE. At Tell Aswad, a Neolithic community dated to between 8700 BCE and 7500 BCE, archaeologists discovered evidence of crude earthen walls the inhabitants built using soil and reeds. Excavators at Tell Aswad also detected evidence of very crude brickmaking. Apparently the inhabitants pushed clots of clay-rich soil into molds built with the common reeds of the area. The bricks then dried in the sun, a lengthy process that can take weeks or months, requiring constant rotation for even results.

Ancient Egyptians used sun-dried mud bricks as building materials. Artwork at the tomb of Thebes shows slaves mixing, tempering, and carrying clay for the sun-dried bricks. In Exodus 5 of the Bible, Moses and Aaron meet with Pharaoh and make their immortal request: "Let my people go!" But instead, Pharaoh instructed his managers: "Ye shall no more give the people straw to make brick, as heretofore: let them go and gather straw for themselves." The bricks made in the time of Moses used very fine clay from the Nile River, and therefore dried extremely slowly. Adding straw allowed the bricks to dry faster and added stability, preventing cracking.

Over time, brickmakers began to experiment and eventually stopped using straw. They learned to coat the surfaces of the brick with waterproofing, and added stabilizers such as lime or cement. They also shortened the drying process by constructing large kilns.

There are several advantages to using mud bricks for construction:

- Energy-efficient in both summer and winter, and ideal for passive solar heating and cooling. Indoor temperatures vary only about 5 degrees between summer and winter, making it naturally cool in summer and warm in winter.
- Environmentally friendly. Low carbon footprint to construct, plus more earthquake-friendly, more fireproof, easy to recycle, and good for dampening sound.
- Inexpensive in raw material and use of skilled labor.

But there are disadvantages as well:

- Labor intensive, with lots of bending, carrying, mixing, and transport.
- Mortgage lenders tend to shy away from bankrolling new home construction using mud bricks.
- Weather restrictions mean building in the hottest months to avoid rain damage to fresh bricks.
- Insects and small rodents tend to prefer these structures unless you use dung in the mix or coat with lime plaster.

Clay cements the bricks together and is a key ingredient. Any clay will work, but there are hazards. The adobe soil of the US Southwest contains abundant to excessive clay, but sandy clay or loam soil can be scarce there. Soil with too much clay produces too many shrinkage cracks in a brick. Soil that is too sandy leads to bricks that crumble easily. If after a soil test you find that your ground has too much or too little clay, you can bring in sand or soil with a different clay content as needed. If your soil contains more than 20 percent clay, you can add extra sand or organic material. You'll need to be scientific about your testing and experimentation, and take careful notes so that you can adjust one variable at a time before you can mass-produce high-quality bricks.

Table 23: Soil Names and Composition Ranges

Soil Name	Percent Sand	Percent Clay	Percent Silt
Loamy sand	70–85	0–15	0–30
Sandy loam	50–70	15–20	0–30
Sandy clay loam	50–70	20–30	0–30

Sand is the second key ingredient, acting as a filler for the clay to bind with. Again, just about any sand will work except common ocean beach sand, because the salt upsets the chemical reaction when the clay dries.

Straw or fiber is the third ingredient. It acts as another binder and prevents shrinkage when drying. A ratio you can use is about 0.5 pound of straw for every cubic foot of mud mixture. In some parts of the world, artisans substitute animal dung, pine needles, wheat chaff—the possibilities are endless.

One of your key decisions is how big to make the bricks, which you settle by constructing forms. The typical size for Spanish colonial bricks was about 16 inches long, 10 inches wide, and 5 inches thick. The mold is often the bottleneck in mass-producing bricks; you can experiment with forms that mold up to six bricks at a time, although you should probably start with single-brick molds during your testing phase. You may find it helpful to dust the insides of the mold with dry charcoal before filling it, or wet it down with water.

You will want some kind of mixing area close to the drying zone where finished bricks will cure. Ancient cultures used dirt pits, where they mixed the resources with their feet.

Here is a simple process for creating mud bricks:

1. Test your soil to determine the clay content, with 20 percent clay the optimal range.
2. Dump the soil into a pit, or over a flat area in a pile.
3. Mix sand into the soil if the clay content is too high. For example, if you have 100 pounds of soil that is close to 50 percent clay, you need about 100 pounds of sand.
4. Add straw, in a ratio of about 0.5 pound of straw for each cubic foot of mix.
5. Mix thoroughly, using a hoe, shovel, cement mixer, the feet of several humans, etc.
6. Judge the mix and determine if it needs adjustment with water if too dry or soil if too wet.
7. Wet the mold with water and place the mold in the area where the bricks will dry.
8. Shove the mix into the mold. Press with your thumbs to ensure no bubbles or voids at the corners. Smooth the top as flat as possible.
9. Remove the mold carefully, lifting straight up. You will soon learn if your mix is too damp (puddling) or too dry (leaving voids).

10. Leave the bricks in place to begin drying.

Avoid direct sunlight when "sun-firing" bricks to prevent rapid drying on one side while still cool on the bottom. After three days you must turn the brick on its side and continue drying for another three days. Generally, the longer you can cure the bricks, the better, but you can run into weather issues in wetter climates.

You can test your finished product with a simple drop test. Allow the brick to fall to the ground from about waist high. If it breaks, your recipe probably has too much sand. If it is already cracked, it has too much clay. Your best bet is to allow time for some practice runs to figure out what kind of soil you have, how much of the sand and straw additive to use, and how long to cure. Once you have the recipe dialed in, you can ramp up production.

Depending on weather conditions, it can take a couple of weeks for mud bricks to dry. But the payoff is that sun-dried bricks can last up to thirty years.

To speed up the brick-forming process, some builders use a "Cinva Ram," a manually-operated compactor that squeezes your mud mix into a brick form. Such bricks require less water, and cement is often added to the recipe for more stability.

Ruins in central Rome near the Appian Way, showing crumbling clay brick structures

Roman brickmakers around the first century BCE began stamping each brick with an identifying logo that they pressed in while the clay was still wet. Before kilns took over, Roman brickmakers might allow up to two years of drying before use. Once they developed mobile kilns, brickmaking took off throughout the Roman Empire.

ABOUT CLAY BRICKS

Clay bricks usually have a ratio of about 4 parts clay to 1 part sand, and the typical size is much smaller than for adobe, usually on the order of 12 inches long, 4 inches wide, and 4 inches thick. There are numerous variations, however. The monarch style brick can reach almost 16 inches long and weigh up to 12.5 pounds. The standard Roman brick was 11⅝ inches long, 3⅝ inches wide, and 1⅝ inches thick. A modern, standard brick weighs 4.5 pounds and is 8 inches long, 3⅝ inches wide, and 2¼ inches thick.

In India's Indus Valley, the "Indus Proportion" for baked bricks dating back to 2600 BCE was set by standard at a ratio of 4:2:1 for length, width, and height. Thus a 12-inch brick would be 6 inches wide and 3 inches thick. Researchers Aurangzeb Khan and Carsten Lemme tie the advance from sun-dried to baked bricks as fueling the rise of major cities throughout the Indus Valley, from the start of brick use around 7000 BCE to the first use of mortar, suggested by archaeologist Roman Ghirshman, at the Mehrgarh of Balochistan in the Indus Valley.

Ancient Egyptians used mud, clay, and sand as a mortar for the early pyramids, and experimented with lime mortar, gypsum mortar, and other ingredients. Gypsum proved too soft, however. The Babylonians experimented with lime and pitch, while the Greeks added volcanic ash and developed a hydraulic cement that could cure underwater, useful for aqueducts. The Romans improved on the methods and created what became known as pozzolanic mortar, based on volcanic ash rich in quartz and aluminum, which reacts with slaked lime to form a very hard and strongly binding product. The name comes from volcanic pumices and tuffs found at Pozzuoli (Naples).

In a 2017 article at Pubs.GeoScienceWorld.org via the *American Mineralogist* magazine, researchers worked out the complex chemistry surrounding Roman marine concrete. Apparently the presence of two obscure minerals, phillipsite (a zeolite with the formula $[(K,Na,Ca)_{1-2}(Si,Al)_8O_{16}\cdot 6H_2O]$) and aluminum-rich tobermorite $(Ca_5Si_6O_{16}(OH)_2\cdot 4H_2O)$

strengthened undersea concrete pilings significantly. These structures still stand after 2,000 years—much longer than typical concrete lasts.

Meanwhile, engineers in China developed a substitute for volcanic ash, which is not widespread there. Instead they used a mix of sticky rice soup and lime.

Making Mortar for Bricks

As mentioned, for adobe bricks you can use the same adobe mud mixture for mortar. Bricklaying is complicated, and if you plan to make your own clay bricks, you'll also need to make mortar. If you can purchase cement, you use a simple recipe of 1 part Portland cement to 3 parts sand. Mix thoroughly and add 1 part water to 3–4 parts mix. If the mortar doesn't have enough water, the bricks won't stick together properly. If it's too wet, the runny mortar will overflow from the joints, which is messy and wasteful.

If you don't have access to cement, you'll have to make it. You can use limestone or seashells that you burn either in a kiln or in a hot fire. Eggshells might work, but you'd have to experiment. Break up the limestone into very small pieces, layer in a stacked wood bonfire, and burn for several hours. Allow to cool and then pull out the crumbly pieces. If you have too many hard pieces left, reburn them.

The crumbly burnt limestone is called quicklime. You need to convert your quicklime to slaked lime by placing it in water, at a ratio of about 4 parts water to 1 part quicklime. Perform this process in a lined pit, a concrete pen, or even a wheelbarrow, but use caution. The process gives off heat and gas, and the mixture may bubble ferociously. The less water you use, the more likely that you'll end up with a wet slaked lime putty, which you can mix with sand to get mortar.

A rough ratio for creating concrete is about 1 part cement, 2 parts sand, and 3 parts gravel, then enough water to create a thick slurry you can pour, or at least form. To make mortar, try 1 part slaked lime putty to 3 parts sand.

By the way, not all sand is created equal, which is why your mileage may vary with any project you start from scratch rather than store-bought material. The best sand to use for mortar has sharp, angular edges. This type of sand comes from washing soil and using gravity separation to clean away organic particles and clay. The sand you get from streams, creeks, and rivers is often not the best for mortar, because those sand grains are almost perfectly round. The higher up a drainage, the better your chances of finding angular sand particles. Long periods of movement in swift water will knock all the sharp edges off sand particles, and once they reach the ocean surf, they might also have salt crystals interspersed, further interfering with the chemistry.

MAKE YOUR OWN CONCRETE BRICKS

Making adobe or clay bricks is fairly low-tech compared to making your own concrete bricks. But mud and adobe don't last nearly as long, and in the case of adobe, you may have to make repairs after severe weather. At wikiHow.com, there is an extensive article on making bricks from concrete. It covers several concepts well: how to construct the right mold to mass-produce concrete bricks, how big to make them, and more. However, that article relies on bagged, commercially available dry concrete mix.

If you have to go all the way to making concrete bricks (or mortar) from scratch, there will be considerable trial and error. Some recipes call for 6 parts sand to 1 part cement. But that ratio can change to 4 parts sand to 1 part cement, to 3 parts sand to 1 part cement for load-bearing walls. At Survival-Manual.com/cement.php there is a good set of instructions for starting completely from scratch. They start with some definitions:

Cement is pulverized limestone, oysters, freshwater mussels or seashells that have been heated to high heat to remove CO_2 (slaked lime).
Concrete is a mixture of cement, water, sand, and gravel.
Mortar is a mixture of cement, water, sand, and lime. Slaked lime can work as mortar, but it's better to stretch it.

Again, your mileage may vary when mixing your own concrete, but the common ratio is 1 part cement (usually slaked lime); 2 parts clean, dry sand; and 3 parts gravel. Mix the dry ingredients first, then add water slowly until the mix is workable and not runny. Start with small batches. Once you get the knack, you can proceed to pouring the mix into molds and drying them in the sun (or a kiln).

The advantages of using concrete for bricks include the following:

- Stronger than mud or clay
- More water-resistant
- Easier to make straight walls and square outlines
- Better able to bear loads

If you can purchase your own cement, you will obviously save considerable time and effort. The most familiar type of cement is Portland cement, named for a similarity with Portland stone, quarried at the Isle of Portland in the English Channel near Dorset, England. Portland cement was patented by Joseph Aspdin in England in 1824, greatly improved by his son William about twenty years later, and further refined until 1859, when the London Sewer Project required strict, uniform material. Today Portland cement is composed of calcium, silica, alumina, and iron. Calcium is derived from limestone, marl, or chalk; silica, alumina, and iron come from the sands, clays, and iron ore mixed in. More specifically, Portland cement contains the main chemicals listed in table 24.

Table 24: Chemicals Found in Portland Cement

Compound	Formula	Percent by Weight
Tricalcium silicate	Ca_3SiO_4	55
Dicalcium silicate (belite)	Ca_2SiO_5	20
Tricalcium aluminate	$Ca_3Al_2O_6$	10
Tetracalcium aluminoferrite	$Ca_4Al_2Fe_2O_{10}$	8
Gypsum	$CaSO_4$ $2H_2O$	5
Potassium oxide	K_2O	Up to 2

Following are some common sources for the main components:

1. Lime (calcium oxide), CaO, derived from limestone, chalk, shells, shale, or calcareous rock
2. Silica (quartz), SiO_2, derived from sand, old bottles, clay, or argillaceous rock
3. Alumina, Al_2O_3, derived from bauxite, recycled aluminum, or aluminum-rich clay
4. Iron, Fe_2O_3, derived from clay, iron ore, scrap iron, and fly ash
5. Gypsum, $CaSO_4$ $2H_20$, derived from evaporite deposits

Typically the ingredients are heated to high temperatures to form a clinker, then ground to a powder. European Standard EN-197-1 stipulates a mixture of 66 percent calcium silicates, with the Ca:Si ratio not less than 2:1.

COB

Cob, sometimes called cobb or clom, is basically a free-form adobe. Cob is a mixture of clay, sand, and straw in the same ratio as adobe bricks, but instead of using a mold, you leave the mixture thick so you can pile it up or slap it onto a framework and allow it to dry in place. It is closely related to the "wattle and daub" technique, which uses a wattle or frame that gets coated with a familiar mix of soil, clay, straw, and/or dung. Since the material you'll use can come directly from the property you build on, cob is extremely cheap to make. You won't battle obnoxious odors, such as outgassing from treated lumber, and the high thermal mass of cob keeps insulation costs down. The thicker the walls, the more energy-efficient your structure.

The use of cob for construction goes back thousands of years, and the technique is still used because it is simple, inexpensive, and fairly easy to master. Cob looks appropriate with stone and timber, so you can boost aesthetics and enhance structural components in countless combinations. Cob tends to be earthquake resistant and resists cracks; and, when properly cured, rain and snow won't degrade the material once it dries into place.

Most of the same cautions for adobe mud bricks apply to cob construction. If you use the right ratio of clay, sand, and straw, a cob house will easily shrug off rain, snow, or hailstorms. The durable material doesn't degrade and will last for years to come. It stays completely dry, and the sound-dampening qualities make a cob house both cool and quiet.

Cob has an advantage over brick regarding organic shapes—anything whimsical, like curved walls, domes, arches, and vaults, just requires some kind of framing. In a sense, you are sculpting with cob.

To create a cob wall, you would start with a wide, fat base, then taper as you go up. Instead of using bricks, you form the walls with your hands. Each layer of cob must be allowed to dry thoroughly before laying the next. Due to their width, cob wall structures lend themselves to having load-bearing walls; however, a wood beam system commonly supports the roof structure. Usually a two-coat earthen or lime plaster serves as a final finish.

RAMMED EARTH

As its name implies, rammed earth construction involves mechanically compressing a mix of sand, gravel, clay, and water into a form. Some techniques add cement into the mix for additional stability. Rammed earth construction dates back thousands of years, and cultures from China to Europe and Africa used the process.

The challenge in using rammed earth is to create a movable form that you can continually raise as the walls take shape. Good construction skills are required to form windows, doorways, and other openings as you shape out the structure. Each "lift" of raw earth is about 6 inches, which is then smashed down to squeeze out air. You can use coloring agents to create interesting patterns.

The exterior walls need some kind of treatment, a wash or even beeswax, to prevent water damage. Like cob houses, some kind of wooden roof structure tops off construction, but such materials can degrade the fire-protection qualities of using soil for the walls. Rammed earth buildings offer impressive thermal protection, guarding against extreme heat and cold.

ROCK WALLS

Rock walls help build up a homestead in a variety of ways. They add landscaping features to terrace hillsides, add protection from predators, provide a safety feature against larger intruders, help funnel water

Dry-stacking angular rocks takes artistry, skill, imagination, and patience. This old wall in northern Nevada is probably over one hundred years old.

and air, and create pits for composting, storage, or rubble. You can build them from adobe bricks or mud, but if you already have a lot of rocks on your property, stacking rocks can serve more than one purpose.

Check the Regulations

Many cities and counties restrict the height of any rock wall to 4 feet, so you'll definitely want to check with neighborhood organizations or governing agencies.

Choose Your Rock

You'll first want to decide whether you are simply dry-stacking the stones or plan to use some kind of mortar to hold them together. The best stones to use for dry-stacking are quarry stones, typically andesite or basalt volcanic rocks, which are jagged and angular with sharp edges and must be pieced together like a jigsaw puzzle. The interlocking edges provide stability when built correctly, but shifting and earthquakes can bring them down.

River rocks are typically smooth and round, ranging in size from fist-sized cobbles to larger stones difficult to move. These rocks require mortar, in the form of either mud or cement. The cost of cement to hold river rocks together can outweigh the cost of the rock, but the rounded

This ancient rock wall at Ephesus in Turkey shows a variety of rocks mortared together.

rocks can look better to the eye. You'd typically use plenty of mortar, then clean with a stiff wire brush before the mortar sets completely to let the stone face shine through.

At TheStoneTrust.org, master craftsman Brian Post has an extensive and informative post about building stone walls. He divides wall stone into two basic categories: level bedded and irregular. "Level bedded stones have parallel top and bottom surfaces, and will often split into thinner stones," Post writes. "Slate and shale and sandstone are typically level bedded stones. Some limestone and schist are also level bedded. Some wallers will also refer to level bedded stone as regular stone." Post asserts that level bedded stone results in a neater wall.

Irregular stone accounts for massive rocks that come in varied sizes and shapes. There may be very few flat parallel surfaces to work with, as well. "Granite and marble both break into irregular shapes. There is a continuous range from stone that is clearly level bedded to stone that is clearly irregular. Most stone is somewhere in between. Irregular stone can also be cut, or split using feathers and wedges, to form regular shapes that it would not form naturally." Post says that irregular stones provide a more rustic look, as do walls with larger stones, while smaller stones tend to look tidier. You may have to work around larger rocks; you also may have to use a hammer to shape an irregular rock into the form you need.

Estimate the Tonnage

Clean, smooth river rocks with a variety of colors make for an interesting mortared wall.

When buying stone by the ton, you can estimate about 10 cubic feet of wall per ton of stone delivered. You'll have to use a calculator to determine decimal fractions if you're not making your wall 1 foot thick, so the math can get tricky. If your wall is going to be a solid 2 feet thick and 45 inches high, 1 ton of stone will give you 2 feet of length. A 20-foot wall would require 10 tons. A good rule of thumb is about 1.8 tons per cubic yard of wall built, but you can round that up to 2 tons/cubic yard to make the math easier and assume that you'll end up with more rock.

Use Solid Foundations

Your foundation is crucial, as you'll learn from the plentiful resources available, including DIY books, YouTube videos, and government pamphlets. Most experts recommend digging down about ½ foot, then completely clearing a running area about 3 feet wide. Remove all roots and organic debris, and tamp down the bare soil to prevent later shifting. Stomping hard on the soil repeatedly with heavy boots usually does the trick, but there are handy tamping tools available. Then liberally add in pea gravel for drainage and proceed with a 2-foot-thick wall.

Start your runs with larger rocks, and make the wall thicker at the base, regardless of whether you use mortar. You'll need a variety of heavy hammers, ranging from 2 to 16 pounds if possible, plus chisels, a string line, shovels, trowels, a tape measure, gloves, goggles, tough pants, heavy boots, and possibly even rebar or wood framing. You'll also need lots of practice, and your first dry-stacked wall may get recycled into another structure. You may want to avoid moving too many "two hander" boulders, as the back strain can add up quickly.

Smooth river rocks in a foundation add an interesting, sturdy appeal.

PATHS AND WALKWAYS

Using stones for paving walkways is an easy way to add function and artistry to a garden landscape. The rocks cut down on mud; make channeling runoff easier; and tie in walls, bridges, and other features.

Slate is fairly easy to work with; the biggest challenge is in preparing the bed where you lay the stones. If you don't take care to smooth out all the bumps and pockets and meticulously tamp down the soil, the thinner stones can snap or fracture. Some artisans use mortar; others prefer to leave soil between the stones and add plants such as Corsican mint, which gives off a heady fragrance when you step on it.

Granite mixed with basalt in a variety of shapes and steps adds visual appeal to this Japanese garden.

ROOT CELLARS

Farms, ranches, and similar homesteads all tend to require multiple buildings. Barns, chicken coops, drying sheds, mills, privies, forges, kilns, maintenance sheds, greenhouses—the list is endless. One of the first secondary structures many homesteaders require is the humble root cellar.

Although they vary greatly in design and use, a root cellar is basically any storage location that takes advantage of the natural cooling and insulating properties of rock and soil to maintain a suitable storage temperature for food stock. In her classic book *Your New Root Cellar* (Atlantic, 2011), Julie Fryer writes that archaeologists have found evidence of ancient root cellars dating to 40,000 years ago in Australia (Fryer, 18).

In seventeenth-century Europe such structures were important additions to farms, and the practice crossed the Atlantic when colonists took up farming in North America. The tiny town of Elliston, in the Canadian province of Newfoundland and Labrador, boasts more than 130 documented root cellars, with over half still in use, and bills itself as the "Root Cellar Capital of the World."

Before refrigeration, settlers built an underground root cellar to store long-lasting items such as carrots, turnips, beets, parsnips, potatoes, and other root vegetables. Today, root cellars have made a comeback on homesteads where electrical refrigeration is impractical. The cellars work year-round. They maintain an even temperature around 45°F (7°C), which keeps food from freezing during the winter and safe during the summer. They also keep fermented spirits safe, provide storage for extra drinking water, and generally enhance food security.

There are many different plans available in Fryer's book and elsewhere; even the US Department of Agriculture has a pamphlet available. You can dig extra space in a home or barn basement or excavate a freestanding cellar. Some key considerations follow:

1. Pick a location somewhat close to the main house for easy access during winter storms. You don't want to risk life and limb just to get a few potatoes.
2. Check the sun, and plan for little exposure on the door itself, unless you are very far north. North or northeastern sites remain coolest, but if you'll need some help from the sun during the winter months to avoid freezing, keep that in mind.
3. Check for drainage, and slope the floor so it drains out the door. You don't want any puddles. Use plenty of gravel, and check the slope around the outside of the structure.
4. Look for existing hillsides or elevated rises that don't risk cave-ins. You can start the structure by digging a little way into the hill, build it out, then use the soil from the hillside for insulating cover.
5. Consider whether you'll need passive light when siting the door. Stringing electricity for light may not be practical, and many root cellars require a flashlight or battery-powered LED lights.
6. Check to see if you need a permit. Most localities do not require a building permit for a root cellar, but some do. If you'll be using a backhoe, make sure you know where your utilities come in from the street.

7. Use on-site materials as much as possible. Fancy structures with framing, insulation, drywall, and concrete are not an option for some homesteads. Stacking rocks, using adobe bricks, and log roofs have worked for thousands of years, and most older root cellars had a packed-dirt floor. You can find plans online that use old tires, earth bags, cinder blocks, homemade bricks, and more.
8. Plan for shelving, and allow for good air circulation.
9. Earth insulation requires a strong roof. You may end up with tons of earth atop your structure, so plan accordingly.
10. Make sure your door fits snugly, allowing for no invasion by insects and other pests. You may want the root cellar to double as a security feature, with a door that locks from the inside.

If you're homesteading on an existing farm or ranch, the ruins of a root cellar may already be present; you'll just need to find and retrofit the structure.

This old root cellar near Bishop, California, had a very low ceiling and probably a door.

Note the different bands of color from this soil test, which is about 45 percent clay, 45 percent sand, and 10 percent silt. The finest, lightest clay is still suspended in the reddish layer at the top, so ignore that. The thick light brown layer below it is good clay. The dark, thick layer of sand at the bottom is almost half of the sample. There is a thin tan silt layer in the middle.

Walkway construction using basalt mini-boulders and store-bought mortar mix

CHAPTER 11

CIVILIZATION

In his book *Sapiens*, Yuval Noah Harari points to a time about 70,000 years ago when *Homo sapiens* "started doing very special things" (Harari, 21). He explains that bands of explorers left Africa a second time, driving Neanderthals from the face of the Earth. In the next 45,000 years, *Homo sapiens* invented boats, oil lamps, bows and arrows, needles for sewing, and even art. "The appearance of new ways of thinking and communicating, between 70,000 and 30,000 years ago, constitutes the Cognitive Revolution," Harari explains. Was it some accidental genetic mutation? Harari calls it "The Tree of Knowledge" mutation, and believes it was the richness of our language that cleared the way to store and share complex thoughts—as well as gossip about our neighbors.

In his Pulitzer Prize–winning book *Guns, Germs, and Steel* (W. W. Norton, 1999), Jared Diamond notes that from the time 7 million years ago when we diverged from great apes, humans subsisted by hunting and foraging. Sometime around 11,000 years ago, humans turned to what may be called food production. We planted and harvested various plants and domesticated certain animals. "[T]he ancient Chinese developed [food production] independently by themselves, while others (including ancient Egyptians) acquired it from neighbors. But . . . food production was indirectly a prerequisite for the development of guns, germs, and steel" (Diamond, 86).

IMPROVING SOIL

You can gain some easy hacks for improving your soil at RuralSprout .com, which lists this advice:

1. Do no harm. Use organic practices; avoid powerful fertilizers, pesticides, and herbicides; and keep the soil healthy using natural practices.
2. Dig or till as little as possible. Every time you send a shovel into the soil, you damage the natural structure. Disturb that structure as little as possible to maintain soil health.
3. Avoid bare soil. Believe it or not, you can actually "sunburn" soil

if you leave it uncovered; the intense rays of the sun damage microbes normally found in darkness. Even the cheapest covers—alfalfa or clover—can serve the purpose.

4. Reduce foot traffic. Your bulk compresses soil and drives out oxygen. Soils need aeration to breathe and supply air to the billions of microbes per spoonsful present.
5. Grow nitrogen fixers such as legumes that fix nitrogen into the soil, such as green peas, kidney beans, garbanzo beans, and peanuts.
6. Rotate crops. If you grow the same crop over and over, you deplete the soil of specific nutrients. Switching from legumes periodically avoids disease too.
7. Add green manures such as chopped-up plant clippings to bring in nutrients.
8. Don't rely on annual plants—mix in perennials to keep seed costs down and allow roots to strengthen.
9. Add compost for soil health. There are brown and black composts, such as rotted animal manure, worm castings, wood chips, and also consider biochar—charcoal produced by heating wood, grass, straw, or other organic waste materials.
10. Avoid runoff by using terracing, landscaping, swales, basins, ponds, basins, or channels.

Because soil science is only a remote cousin of geology, we'll leave the complex discussion of soil types and improvement to the volumes of material published by the US Department of Agriculture, the Soil Conservation Service, the US Geological Survey, and other organizations. What matters is that you understand that soil is a living, breathing thing. When soil is healthy, there is a larger population of microbes in 3 tablespoons of healthy soil than there are humans on the entire Earth—some 8 billion at latest count. The decomposition of organic matter is a key component, feeding the various organisms.

In general, the darker the soil the better, but there are exceptions. Your best bet is to test your soil for pH; know your percentage of sand, silt, and clay; and rotate crops to fix nitrogen and avoid depleting key nutrients. You can spread livestock manure, mix in straw and the ash from wood fires, and use various other methods to conserve or improve soil.

Getting back to geology, you'll do well to remove as many of the larger rocks as possible; many farmers pile rocks at the margins of their fields both as boundaries and to serve as inventory for homestead construction projects. In addition, as you pick up stray rocks, you'll learn more about the geology underfoot. You can note the degree of roundness, which gives you an idea of how far rocks might have traveled, or whether they are simply rubble from the bedrock below that worked its way to the surface. Take the time to determine whether the rocks are igneous, metamorphic, or sedimentary, and attempt to further narrow down their families.

Note, however, that there are various methods for bringing in stray rocks from hundreds, if not thousands, of miles away. In Oregon's Willamette Valley, dozens of feet of fertile topsoil resulted in part from the repeated inundation of the valley from the Missoula floods, originating in Montana when the Clark Fork burst through an ice dam and scoured its way to the ocean. The ice carried boulders and cobbles of considerable size and deposited them as erratics miles from their origin, placing granites, schists, and gneisses where only volcanic and sedimentary rocks should rest.

Night Soil

Throughout history, struggling homesteaders seeking to improve their soil have employed "night soil" to add nutrients. It's a nasty process, fraught with danger from human-borne pathogens and parasites, and smelly to boot. Its use dates to ancient societies in African, Central America, Greece, China, and elsewhere and is still widespread in some developing countries.

GEOLOGY AND MAJOR CONSTRUCTION

The link between geology and major construction projects is crucial. To build something large enough and solid enough to stand the test of time, our forebears used rocks. And not just any rocks—they quarried, carried, and constructed with the best materials at hand. For example, the Incan engineers who constructed the high-elevation complex at Machu Picchu took advantage of a graben—a sunken block lying between two faults. They quarried the Quillabamba granite, a 246-million-year-old intrusion, from an outcrop on-site, and with such precision that they didn't need mortar.

The following segments are not meant to be exhaustive, or to exclude any culture or nationality. These examples simply represent important structures showing several different applications of practical geology.

Blueschist Structure: Stonehenge

The famed blueschist stone circle at Stonehenge in England made use of a rock with little obvious value. Blueschist, a glaucophane schist, is a metamorphosed basalt, cooked at high pressure to between 200°C and 500°C (392°F and 932°F). It turned blue from the presence of the minerals glaucophane and lawsonite. It is hard, dense, and strong, and perfect for longevity.

Ancient artisans constructed Stonehenge from 3000 BCE to 2000 BCE, apparently as a cremation burial ground featuring a giant solar observatory, perfectly aligned for marking the summer equinox. The bluestones originated in the Preseli Hills of southwestern Wales, about 140 miles from Stonehenge. The larger sarsens are silicified sandstone boulders found in West Woods, about 16 miles from the monument. The sandstone altar stone is probably from east Wales.

One facet of the blueschist that may have set it apart was its acoustic qualities. When struck, it rings in a solid, pleasing tone. Such rocks are called lithophones, and this could also explain why the rocks were prized enough to transport using sleds, logs, or some other means.

There are an estimated 1,300 stone circles in Britain, Ireland, and Brittany. There is evidence of hundreds of stone circles in Europe, starting as early as 5,000 BCE in northwestern France. For more information,

Stonehenge, a Neolithic stone monument constructed from 3000 to 2000 BCE GARRET WISCOMBE.

search for "The Megalith Map" by Aubrey Burl, accessible via The Wayback Machine. Or check out *The Significance of Monument: On the Shaping of Human Experience in Neolithic and Bronze Age Europe* (Routledge, 1998) by Richard Bradley.

Limestone Structure: The Great Pyramid

The Pyramid of Djoser, built c. 2630–2610 BCE during the Third Dynasty, is generally considered to be the world's oldest monument constructed of dressed masonry. The Great Pyramid of Giza, built by Khufu in the early 2500s BCE, is the largest Egyptian pyramid and is the only remaining structure of those considered the Seven Wonders of the Ancient World.

The Great Pyramid contains 6.5 million tons of stone, composed of 2.3 million limestone blocks. Each limestone block, quarried from a nearby limestone deposit, weighed on average 1.8 tons. The limestone blocks are mostly crude cubes or rectangles, and show little evidence of precision "dressing" or shaping. The external blocks are mortared together. The outside of the Great Pyramid is clad with a brilliant white limestone from nearby Tula.

Inside, pink granite blocks from Aswan made up the King's Chamber. These blocks are precisely shaped and fit perfectly. The sarcophagus is a red granite.

So far, 118 Egyptian pyramids have been identified. Many are just mounds of rubble of ruined mud brick; some of the lesser pyramids used inferior grades of limestone, so also tumbled into ruins. Archaeologists believe the pyramid of Neferirkare Kakai started as a step pyramid; engineers then converted it to a "true" pyramid by filling in the steps with loose masonry. The Red Pyramid is the world's first successfully completed smooth-sided pyramid. The structure is also the third-largest pyramid in Egypt, after the pyramids of Khufu and Khafra at Giza.

At the southern end of the "pyramid belt," engineers saved time and effort by incorporating part of a natural limestone hill in their design. Engineers later incorporated this concept when they constructed the Sphinx, carving it directly from a limestone outcrop.

Researchers believe engineers floated the large limestone blocks down the Nile, then dragged them via sleds in canals of wet sand. In 2013 archaeologists discovered the "Diary of Merer," written on papyrus by an Egyptian official who documented several processes.

Marble Structure: The Parthenon

The Parthenon, dedicated to the goddess Athena, is situated on the Athenian Acropolis. It is the finest example of ancient Greek architecture, and the most perfect Doric temple ever built. Workers began construction of the temple in 447 BCE, when Athens was at the peak of its power. Eleven years later they finished—an astounding feat. The temple has survived numerous sackings, lootings, and invasions, and over recent years Greek authorities have devoted considerable energy and expense to restoring the structure.

Callan Bently, a geology instructor at Piedmont Virginia Community College, has worked out the complex geology beneath the Parthenon. In his award-winning *Mountain Beltway* blog for the American Geophysical Union (blogs.agu.org), he wrote an amazing and easy-to-understand article titled "Geology of the Acropolis" in 2015. He showed how the cap of the hill where engineers built the Parthenon is a *kilippe* (CLIP-uh)—a German term for an erosional remnant of a thrust sheet. The Tourkovounia Limestone is about 90–100 million years old, but it sits *atop* the Athens Schist, which is no older than 72 million years. That directly violates the Law of Superposition, postulated by Nicholas Steno in 1669, that older rocks are always at the bottom.

Numerous resources went into the Parthenon. Egypt supplied ivory for the statues, Thrace and Libya supplied the gold adornments, and the quarry at Penteli supplied the marble. Pentelic marble has a unique

The Parthenon in Athens features Pentelic marble, which glows with a golden aura in sunlight.

white appearance that shines with a golden hue in sunlight. It has excellent purity, persistent clarity, and is nearly transparent in thinner sections. In fact, high-quality marble from Mount Pentelicus was an important export for ancient Athens, spawning numerous other great monuments around Greece. The main quarry is now protected by law, reserved only for the Acropolis Restoration Project.

Skilled quarrymen cut the blocks of marble for the Parthenon in precise shapes, avoiding any faults in the marble. They used iron tools—picks, points, punches, chisels, and drills. According to an article at MiningGreece.com, "One of the secrets of the construction was the metallurgy of the stone masonry tools and the highly skilled stonemasons. Based on the quality of the traces left on the marbles, it appears that their tools were much better than today's tools. It is evident that they had come to some remarkable metallurgical recipes at that time, following a very rigorous experimental research. These recipes were lost, like other special knowledge, along with the decline of the ancient world."

Travertine Structure: The Colosseum

Construction of the Roman Colosseum was completed under the Emperor Titus in 80 CE. A lasting icon of the strength of the Roman Empire, it has withstood numerous earthquakes and periodic looting

The exterior of the Roman Colosseum used 3.5 million cubic feet of travertine.

of its stones. At its zenith it could seat as many as 80,000 spectators, watching gladiators and other events. The Colosseum was the first famous architectural use of travertine, and was the largest amphitheater in the world at the time.

Travertine is a type of limestone that is formed by mineral-rich hot or cold springs. The rapid precipitation of calcium carbonate, or calcite, combined with other minerals, creates the unique swirls and waves that give travertine its distinctive character. Its porous cousin, tufa, is noted in columns at Mono Lake, California, and in huge slabs around hot springs at Yellowstone National Park.

Engineers built the exterior of the Colosseum using 3.5 million cubic feet of travertine, mined from the famed quarries at Tivoli, about 16 miles east of Rome. According to TexasTravertine.com, the builders substituted 300 tons of iron clamps for mortar.

Sandstone Structure: Mesa Verde and Angkor Wat

One of the benefits of sandstone as a construction material is that you can shape it fairly fast with simple tools. Consider the UNESCO World Heritage Site at Mesa Verde in southwestern Colorado. With more than 5,000 sites, including 600 cave dwellings, and occupying more than 50,000 acres, it is the largest archaeological preserve in the world. And it was built without the use of metals.

At Mesa Verde, beginning about 1190 CE, ancestral Pueblo people began building structures from sandstone directly in and against the overhanging walls. The famed Cliff Palace held 150 rooms, 23 kivas, and probably housed more than 100 people. They used an assortment of hammerstones, stone axes, and sharp knives, using cobbles from the local rivers and streams to neatly trim and fashion rectangular bricks. They plastered the bricks using adobe mortar.

The windows at Mesa Verde are functional, usually face south, and the towers were likely situated for lookouts to scan the surroundings for enemies. Pit-dwellings and kivas were clever and well designed to allow airflow for wood fires. Artisans fashioned pottery in kilns, producing pitchers, ladles, bowls, jars, mugs, and platters. Their engineers fashioned intricate systems of silt-retaining dams and reservoirs to husband the precious rain. Famers grew corn, beans, squash, and gourds, and hunters prowled the countryside for turkey, deer, elk, and other large mammals.

Close-up of the cliff dwellings at Mesa Verde, showing rough sandstone construction in the base, and regular formed bricks for the structures

Another world-famous sandstone structure is the temple complex at Angkor Wat in Cambodia. It was erected in the tenth century CE by the first kings of the Khmer Empire out of an extremely silica-rich sandstone quarried from the Kulen Mountains. Had the builders used marble or limestone, the relentless encroachment of the jungle would have reduced Angkor Wat to rubble in short order, but silica, or quartz, is much more resistant to acidic forces.

Mixed Structure: The Great Wall of China

To repel invaders from the Eurasian steppes, control immigration, and collect trade duties, Chinese emperors set about constructing a series of fortifications beginning as early as the seventh century BCE. In the very beginning, villagers simply piled up rocks or used earth, stone, sand, lime, and wood. They also developed a rammed-earth system, using forms to hold earth, then compressing the soil. In truth, they used whatever was at hand, and employed plentiful human labor. They eventually created strengthened foundations using granite, marble, limestone, and sandstone.

During the Ming dynasty, rammed earth and mud gave way to bricks, tiles, lime, and stone. Bricks were far easier to work with, and construction pace quickened. Stones cut in rectangular shapes made up the foundations, gateways, and walls. Evenly spaced battlements, parapets, and watchtowers provided additional protection. Engineers built signal towers at high points, while wooden gates controlled traffic. Builders also constructed barracks, stables, and armories at critical points nearby.

Eventually the Great Wall stretched 12,000 miles, even crossing rivers. Seeking to minimize transportation costs for resources, builders concentrated on the material at hand, using granite in the mountains and rammed earth in the valleys. While some key areas remain, and have in fact been rebuilt for tourists, most of the ancient walls were basically dikes and have long since eroded

The sections of the Great Wall around Beijing, named the Badaling Great Wall, had seen about 10 million visitors per year by the end of 2019. Tourists here marvel at the construction, size, steepness, and view.

Adobe Structure: Taos Pueblo

The Taos Pueblo is a multistoried residential complex representing the longest continually occupied structure in North America. Humans first occupied the area perhaps 9,000 years ago, and there is evidence of

The main structures of the Taos Pueblo are believed to be more than 1,000 years old. KAROL M. THIS FILE IS LICENSED UNDER THE CREATIVE COMMONS ATTRIBUTION-SHARE ALIKE 3.0 UNPORTED LICENSE, EN.WIKIPEDIA.ORG/WIKI/EN:CREATIVE_COMMONS.

considerable trade and travel through there. The remains of its houses and room blocks near here point to permanent dwellings as early as 900 CE, and construction of the pueblo itself probably began sometime between 1350 and 1450 CE.

The original dwellings had few windows or doors, so access was via ladders from roofs made of grass, mud, branches, and plaster. In the beginning builders formed large balls of adobe clay and packed them together. When the Spaniard Francisco Vázquez de Coronado y Luján arrived in 1540, he introduced brickmaking and showed the inhabitants how to build stronger roofs.

Builders used sand, straw, clay, and water to erect their walls, which were thicker and more substantial at the base. The roof used a system of wooden beams called vigas to hold a network of branches and packed soil. The outer walls are continuously maintained by replastering with thick layers of adobe mud, while the interiors are coated with thin washes of white gypsum-rich plaster.

RESOURCE WARS AND CONFLICT DIAMONDS

One of the key challenges for expanding civilizations is continued access to natural resources. The Roman Empire expanded partly in search for more resources such as salt, but also for tin in Britain, a key ingredient in bronze weapons. The War of the Pacific (1879–1884) was a fierce battle between Chile and its neighbors Bolivia and Peru over the rich trade in guano, a key ingredient in fertilizer. The origins of World War I are often blamed on nationalistic antagonism, military buildup, and the assassination of Archduke Ferdinand. But according to the World History Project, behind it all was the competition for African and Asian resources.

In a 2002 article at NewScientist.com, Fred Pearce describes multiple ongoing conflicts either being fought over—or funded by—various lucrative natural resources. Examples cited by the Worldwatch Institute include:

- Diamond mines in Sierra Leone and Angola make the two African nations ripe for plunder by warlords.
- Sapphires, rubies, and timber armed the Khmer Rouge in their interminable jungle war in Cambodia.
- Oil companies prospecting in Colombia encouraged guerillas, using the threat of sabotage to extort hundreds of millions of dollars.

- Opium farming has funded years of war in Afghanistan.
- Columbium and tantalum (coltan), vital resources in the manufacture of mobile phones, funds the Congo's continuing civil war.

During World War II, the German army tried desperately to control oil production in the Middle East and Ukraine. The Soviet Union invaded Finland to secure the nickel deposit in Petsamo. Japan struck at Pearl Harbor in part to access commodities in southern Asia. And the list goes on, particularly regarding petroleum reserves—look no further than Iraq's 1990 invasion of Kuwait.

Still, no continent better exemplifies the constant struggle over geological resources than Africa. At any given time, warlords, armed gangs, and corrupt government officials are fighting over such valuable commodities as diamonds, oil, gold, and rare earth elements. At UN.org, Ernest Harsch described the many conflicts over Africa's diamond trade. In 2005 two non-governmental organizations (NGOs), Global Witness and Partnership Africa Canada, joined with two diamond companies, De Beers and the Rapaport Group, to launch the Diamond Development Initiative (DDI). Its aim was to "promote better conditions for Africa's approximately 1 million small-scale artisanal diamond miners, many of whom earn as little as $1 a day" despite their hard and dangerous work.

Progress has been halting, and Harsch points to the two-edged sword of Africa's bounty: "[T]wo broad concepts [exist]: that the very presence of valuable natural resources often serves as a 'curse' that generates conflict, and that insurgent forces frequently are motivated less by genuine grievances than by 'greed' for money from the control of natural wealth."

CIVILIZATION AND EARLY ART

One sure sign a civilization is thriving, and not just surviving, is the emergence of a vibrant artistic community. The theory is that once a community has solved the challenge of growing and storing enough to eat, its people can devote more energy into honoring the gods, asking the gods for their help, decorating their homes, and adorning themselves with fine jewelry.

According to AncientHistoryLists.com, the ten oldest artworks ever discovered are dominated by fertility fetishes, animal depictions, and strange symbols such as cupules—hemispheric indentations created by percussion on a flat surface. The dates shown in table 25 are often debated, and some are in doubt. For now, note the intersection of practical geology and ancient artistry.

Table 25: Oldest Known Artwork

Name	Age	Location	Medium	Description
The Venus of Fels	38,000–33,000 BCE	Germany	Ivory carving	Fertility
Lion Man of Hohlenstein-Stadel	38,000 BCE	Germany	Wood carving	Man-Animal
El Castillo Cave Art	39,000 BCE	Spain	Ochre	Signs, stencils
Sulawesi Cave Art	45,500 BCE	Indonesia	Ochre	Hand stencil
La Ferrassie Petroglyphs	60,000 BCE	France	Cupules	Fertility
Diepkloof Rock Shelter	60,000 BCE	South Africa	Egg shells	Abstract engravings
Blombos Cave	73,000 BCE	South Africa	Ochre	Earliest drawing
Venus of Tan-Tan	200,000–500,000 BCE	Morocco	Quartzite	Fertility
Venus of Berekhat Ram	230,000–700,000 BCE	Israel/Syria border	Scoria	Fertility
Bhimbetka and Daraki-Chattan	290,000–700,000 BCE	India	Quartzite	Abstract engravings

Note: Some scholars are convinced that the Venus of Berekhat Ram is simply an interesting pebble, and there is skepticism about the Venus of Tan-Tan as well. Dating techniques can be troublesome in the range beyond carbon 14, so interpreting the age of cupules is also challenging. Still, there is growing, if begrudging, acceptance for the general dates at the earliest points of the spectrum. The work of Robert G. Bednarik is especially noteworthy; his 2003 paper, *The Earliest Evidence of Paleoart*, contains considerable detail, citing studies by fellow archaeologists, and is a must-read.

EASTER ISLAND

One of the more notable uses of local resources is the collection of stoic figures called moai (MO-aye) at Easter Island. Archaeologists believe each carved statue honors a deceased family member. Carved using basalt tools from local volcanic ash called tuff between 1100 and 1690 CE, they were hauled into place at considerable effort. One theory gaining support is that the islanders used ropes and "rocked" the statues down the hill from the quarry by shifting the weight from side to side and pulling forward slightly, in a kind of walking pattern.

The Rapa Nui people carved about 900 of the stone monuments, but some never left the main quarry. The tallest statue was 33 feet tall, and the heaviest weighed 84.6 tons. Researchers discovered that many incomplete monuments at the quarry held inclusions of harder stone that the carvers couldn't go around, so they simply stopped and began again on new material.

EARTH PIGMENTS

The oldest-known mine on archaeological record is the Ngwenya Mine in Swaziland, about 43,000 years old. At that site, Paleolithic humans mined hematite to make the red pigment ochre.

Ochre is a kind of clay with ferric oxide (Fe_2O_3) that comes in shades of yellow gold, purple, and brown, but mostly red. "Umber" is the term for material similar to ochre, but umber has more manganese oxide (in a range from MnO to Mn_2O_7)—up to 20 percent—and is thus much darker than common red ochre. Sienna has manganese oxide and goethite, limonite, or jarosite and in color lies between ochre and umber.

Ancient humans used ochre in a variety of ways, in part because it can be such a plentiful resource. There is evidence in Africa that humans used ochre pigments possibly as far as 300,000 years back. It served as body paint for religious ceremonies or, when applied to the entire body, to psych out opponents in battle. Ochre also works as a sunscreen to prevent sunburn. Ochre had many industrial uses, such as to dye animal skins and paint boats. Artisans used ochre as a medium for cave paintings, such as at the caves in France and Spain dating to 25,000 BCE. The Greeks, Romans, Egyptians, Australian aboriginals, Chinese, Indus Valley Indians, and North American tribes all used red, yellow, and brown ochre.

Hillside of vivid red ochre, ready to collect and pulverize

In a 2018 paper, *Mineral Pigments in Archaeology: Their Analysis and the Range of Available Materials*, Ruth Siddall wrote: "[T]he procurement, preparation and application of ochres is arguably the earliest exploitation of Earth materials and the burning of ochres to modify their colour is the earliest form of pyrotechnology."

Collecting and processing ochre is extremely simple: Pick up the rocks, grind them to powder, and use as-is or mix with water, fish oil, or some other medium.

Gypsum salts were a common source of white pigments, as was pulverized calcite. Siddall indicates that huntite ($Mg_3Ca[CO_3]_4$) served as a white pigment for at least four millennia in the Middle East, while Australian aboriginals used China clay (kaolinite). The lead sulfate mineral anglesite ($PbSO_4$) was used as far back as the seventh century CE.

Other mineral pigments included charcoal and manganese oxide for black and the copper minerals malachite for green and azurite for blue. Some clays rich in celadonite and glauconite also served as green pigments. "Blue ochre" among Pacific Northwest Indian tribes contained vivianite, an iron phosphate with the formula ($Fe^{2+}{}_3[PO_4]_2 \cdot 8H_2O$). Siddall goes on to list dozens of other minerals used for coloring and pigments in a variety of ornamental, industrial, and personal usages.

The downside of using mineral and earth pigments for beauty is that the skin absorbs chemicals quickly. Kohl, used by Egyptians as black eyeliner, contains soot, fats, and metals such as lead, antimony, manganese, and/or copper. The men and women of ancient Greece used a white lead facial cream. The infamous "dead white" look on European aristocrats from roughly 1400 to 1800 came from a mix of white lead and vinegar.

Regular use of lead in cosmetics has alarming side effects, including fatigue, weight loss, nausea, headaches, muscle atrophy, and paralysis. Other medicinals and cosmetic products contained arsenic, mercury, carbolic acid, and mercuric chloride, to name a few potentially harmful materials. Fortunately, government agencies began cracking down on harmful additives, but the battle is ongoing.

STATUES, FIGURES, AND OBELISKS

From stacking rocks to carving marble statues, geology has long been important for a range of monuments, from stone cairns to Michelangelo's *David*. Early pathfinders learned to stack rocks as trail markings or erected stone cairns at mountain passes. The Inuit took stone-stacking to a whole new level, creating mythical standing human forms called *inuksuk*. Because the northern tundra has few natural landmarks, these stone constructions worked as navigation aids, markers for food caches, barriers for caribou herding, and more. There could be a single standing stone or a series of large rocks, hefted into place and carefully balanced.

Inuksuk pattern ready for use, made from obsidian needles dug at Davis Creek in northern California

Simply stacking stones is another ancient tradition, involving patience, balance, and good reflexes. Alternately referred to as cairns, stacked stones, "Stone Johnnies," schmoos, and several other names, there are as many reasons for stacking stones as there are practitioners. Some believe that stacking stones has meditative and mindfulness benefits. Others see it as artistic creativity. Still others knock them down and scatter the rocks, seeing such stacks as desecrations and selfish intrusions on the landscape. (***Note:*** Stacking stones is prohibited in national parks and many other public lands.) Your mileage may vary, but certainly your own home garden is a perfect place to erect a primitive "earthquake detector."

Stacked stones in the wild are increasingly common, inviting controversy over forcing human imprints on nature.

PAINTED ROCKS

There is a growing segment of rock art that involves applying paint to smooth pebbles or small cobbles—rock painting. The allure is easy enough to understand: Collecting the rocks yourself is fun, or you can purchase them already sorted. The stone surface has to be smooth enough to hold a good paint, but the "canvas" is small and doesn't take much paint to fix up. All kinds of patterns, writing, imagery, and encouragement are possible—painted rocks are usually positive as well as pretty.

Painting and etching on smooth pebbles is growing in popularity.

BEADS AND JEWELRY

Stones, crystals, and fossils have long been used in personal jewelry. From common agates to expertly faceted diamonds, humans have long sought to turn mineral specimens into showpieces in rings, bracelets, amulets, necklaces, earrings, nose rings, headdresses, and crowns. From common clay beads to the British Crown Jewels, gemstones have been a favorite work material, set in gold, silver, platinum, or other metalwork.

Beads are thought to be the earliest form of jewelry for trade, with materials at Blombos Cave in South Africa dated to 72,000 years old and at Ksar Akil in Lebanon to about 40,000 years old. These were marine shell beads. In his 2015 article "Why We Speak," published in *The Atlantic*, Mark Pagel discusses apparent ancient trade activity at Oued Djebbana in Algeria. He reasons that the discovery of seashells in a location 120 miles inland, in a locale positively dated as containing artifacts 120,000 years old, may in fact indicate that a form of language grew up there. It takes great understanding between both parties to hammer out a deal, after all. Value, scarcity, provenance, potential uses, and a medium of exchange all have to be negotiated. Only a complex language can handle those concepts.

Fossil dentalium have a soft sandstone interior that Native Americans painstakingly picked out to convert the fossils into beads.

GEOGLYPHS

One of the more interesting forms of geological art is the geoglyph. Defined as a large design or motif produced on the ground with durable objects such as rocks, gravel, or even mounded earth, such installations can be up to 8,000 years old, although that estimate is controversial. Sizes range from a few hundred meters to immense. Their distribution across the globe is impressive, with constructions in South America, North America, Europe, Australia, India, Asia, and elsewhere. Modern forms are called "land art."

In 2015 archaeologists rediscovered a 2,000-year-old geoglyph of a killer whale (orca) on a hillside in Peru. At 230 feet long, it was created in negative relief by removing a thin layer of stones to form the outline, then stacking stones for the eyes and other features.

The Nazca Lines in Peru may date back to 500 BCE. Similar to the Palpa orca, they were built by digging out dirt and pebbles to leave a different color of soil exposed. Some of the Nazca Lines are very intricate, consisting of a single line winding around to form a dog, cat, bird, monkey, lizard, even a hummingbird. Other lines are simple geometric shapes. Their purpose is unknown; they are visible from nearby hills, but could be a signal to heavenly gods or meant to mimic constellations.

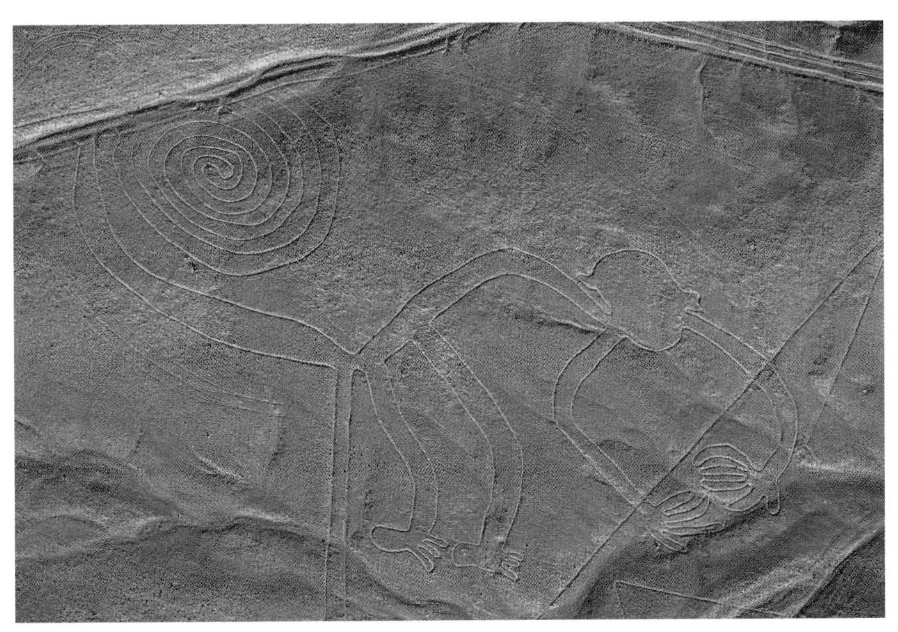

Aerial view of the "Monkey," one of the most famous Nazca Lines PHOTO BY DIEGO DELSO.

One famous geoglyph is the Uffington White Horse, created in Great Britain perhaps 3,000 years ago. Its designers dug trenches and filled them with crushed white chalk, which must be periodically cleared and cleaned. Most experts see a horse in the design, but others imagine it as a large cat. It's an abstract form, stylized and sophisticated, showing the artist wasn't seeking to create an anatomically correct form but instead to evoke a spirit and feeling.

The Boha geoglyphs in India are a series of concentric and linear structures known as the Great Indian Desert Lines. They cover about 1 million square meters (247 acres), and may be as young as 150 years old. Some researchers dispute that they are actual geoglyphs.

Whatever the format or motif, it's clear that geology has played a huge part in the advance of civilization. Rarity, beauty, and whimsy all played a part. Spectacular fossils continually drew the attention of the ancients. In her book *The First Fossil Hunters: Paleontology in Greek and Roman Times* (Princeton Paperbacks, 2000) Adrienne Mayor showed how ancient legends often explained the presence of fantastic fossils by incorporating them into tales of mythical beasts and powerful giants. The ancients fashioned jewelry from shark teeth and ammonites that invited discussions about time, nature, and life itself.

CHAPTER 12
BACK TO THE LAND AND BEYOND

Our Stone Age ancestors survived while worshipping a wide variety of powerful nature gods. Their spiritual skies were filled with deities who orchestrated weather and harvests, managed fertility and the hunt, and thus controlled their lives—albeit capriciously, and often seemingly on a mirthful whim.

As hunter-gatherers, humans learned the downside to overkilling animal stocks, over-gathering seeds and fruits, over-stripping natural resources, and unwisely tipping the balance of the ecosystem. Moving on when resources ran out was not always the best option; bountiful lands nearby might already be occupied by competing bands or tribes. The grueling journey over mountain ranges and across rivers and deserts could be dangerous for the old and weak. In this sense, practical geology already played a huge part in early human life, providing both opportunity and menace and offering lessons in resource management.

Later, when wanderers finally stopped long enough to cultivate, ferment, and otherwise enjoy the fruits of their labors, religions became focused on a single overseeing deity and gradually weaned themselves away from the idea of nature as a force to be feared.

"Ecophilia," from the ancient Greek *oikos*, for household, and *philia*, for love, is the suggestion that humans possess an innate tendency to connect with nature. It was a major tenet of nineteenth-century European geography that humans are shaped by the components of their surroundings. This idea, called environmental determinism, explains why early societies were controlled by climate, forced ever southward by glaciers, and tempted by the bountiful yet dangerous oceans for food.

OUT OF THE GARDEN

The tales of the Old Testament are fraught with exhortations to conquer the wilderness, to subdue it, and to sow the seeds of civilization within it, that all Earth would one day become a garden again. In Genesis, God created the Garden of Eden for Adam and Eve, and the Garden was surrounded by wilderness, chaos, and disorder. The implied duty of civilized humans was therefore to conquer that chaos, extend the order of the original garden, and transform their planet into a worldwide utopia, to go forth and multiply and subdue disorder.

From the book of Exodus comes the story of Moses leading his flock away from the bountiful valley of the Nile and into the forbidding Sinai Desert. Wandering in that blast furnace of doubt and despair, the faith and spirituality of those outcasts was tested over and over again. Yet their resolve, tempered in the chaos, strengthened them so that instead of a small cult lost to history, their story has lived for thousands of years.

In his 1998 book, *The Gifts of the Jews*, Thomas Cahill wrote: "It would be hard to conjure up a landscape more likely to lead to death—a land bereft of all comfort, an earth of so few trees and plants that one may walk for hours without seeing a wisp of green. . . . But this desert brings not death but epiphany, the wildest, most exhausting, most terrifying epiphany of the whole Bible" (Cahill, 133).

Cahill points out that the Jews, who he describes as "a tribe of desert nomads," changed the world with their transition from idols, icons, and fertility symbols to monotheism. This power of the wilderness experience to strengthen and purify became a leading tenet of Christian faith. But a funny thing happened to wilderness as the population grew and the Earth seemed to shrink. There was eventually less to tame and subdue. As we have transformed our planet into the garden we were evicted from, the wilderness that once was outside the garden has now shrunk to become small, isolated pockets *on the inside*. And as we have reached the end of the cycle, the chaos and cleanliness of the wilderness has become both highly sought and subsequently fragile and difficult to preserve.

George H. Williams has shown in his book *Wilderness and Paradise in Christian Thought* how wilderness underwent a transition to become a historically and aesthetically positive meaning. The wilderness motif becomes a metaphor for the struggle between the religion of the wilderness and the religion of the city—or, again, of the fall from the garden (paradise) devastated by sin, wandering in the wilderness, and the vision of a second Eden.

Many gardens constructed during the Middle Ages featured the wilderness/paradise motif, with an Eden-like setting surrounded by chaos. Structure, uniformity, order, and purpose were dominant themes in European and Chinese landscape architecture. The Japanese built a casual, "structured chaos" into many of their gardens. The seemingly offhand chaos was actually quite structured and meaningful.

For Americans especially, there has always been a love-hate relationship with the forces of nature. Early European pilgrims looking westward from their toehold on the Atlantic coast saw matchless tracts of pristine forest, plentiful wild game, and seemingly endless fertile farmland. Yet that abundance was housed in a chaotic, deadly landscape that tested those who would settle it—and theoretically made stronger those who won.

One of the main tenets of the European exodus into North America from the tenuous landing at Plymouth Rock in 1620 had a strongly religious motif. Settlers could flee religious intolerance, grinding poverty, locked-in caste status, and poor future prospects by hacking a productive life and a better chance out of the New World. Perry Miller described the religious zeal behind the settlement of North America in his book *Errand into the Wilderness* (Belknap Press, 1956).

America, Miller wrote, was everything crowded Europe did not have but ardently required: opportunity, abundance, and available land. There was considerable attention given to the notion of starting over—developing new utopian settlements in the untrammeled forests where religious zeal and homestead labor could be meshed in a perfect union before God. It was as though the immigrants had all read Milton's *Paradise Lost*, which includes this passage:

> Christ the Judge will first dissolve
> Satan with his perverted World, then raise
> From the conflagrant mass, purg'd and refin'd,
> New Heavens, New Earth, Ages of Endless date
> Founded in righteousness and peace and love,
> To bring forth fruits Joy and Eternal Bliss.

Conquering this Last Wilderness was essentially a question then of turning the continental chaos into an abundant garden. Farming, the most common enterprise of the times, became a process of clearing the forest, mining the soil, and thus conquering the land. By the early nineteenth century in America, general consensus held that the wilderness was ours to develop, our Manifest Destiny.

Back then, wilderness was at first seen as valueless, something to be eliminated or completely subjugated. "The destiny of the earth is to be subject to man. There can be . . . no real satisfaction to humanity if large portions of the earth remain beyond his highest control," Mormon hierarch John Widtsoe wrote.

James Brooks, writing in *The Knickerbocker* in 1835, said that we do not have to reconcile ourselves to being forever a rude, Philistine order. In the future we shall vindicate our culture. For this confidence we have the highest authority:

> God has promised us a renowned existence, if we will but deserve it. He speaks this promise in the sublimity of Nature. It resounds all along the crags of the Alleghenies. It is uttered in the thunder of the Niagara. It is heard in the roar of two oceans, from the great Pacific to the rocky ramparts of the Bay of Fundy. His finger has written it in the broad expanse of our Inland Seas, and traced it out by the mighty Father of Waters! Oh! that we may consecrate it to LIBERTY and CONCORD, and be found fit worshippers within its holy walls!

Early agricultural practices on wilderness homesteads emphasized complete self-sufficiency, which is generally accomplished by efficient, wasteless utilization of raw products. The independence of the homesteader, the trapper, and the pioneer was to be celebrated and encouraged.

When French sociologist and political theorist Alexis de Tocqueville published *Democracy in America* in 1835, he was struck again and again by the resolve and tenacity of the American farmer:

> It is difficult to describe the rapacity with which the American rushes forward to secure the immense booty which fortune proffers to him. In the pursuit he fearlessly braves the arrow of the Indian and the distempers of the forest; he is unimpressed by the silence of the woods; the approach of beasts of prey does not disturb him; for he is goaded onwards by a passion more intense than the love of life. Before him lies a boundless continent, and he urges onwards as if time pressed, and he was afraid of finding no room for his exertions. (de Tocqueville, 324)

De Tocqueville traveled widely in the United States and wrote admiringly of the homesteader. He saw in those hardy souls a determination that greatly impressed him:

As soon as the pioneer arrives upon the spot which is to serve him for a retreat, he fells a few trees and builds a log house. Nothing can offer a more miserable aspect than these isolated dwellings. The traveler who approaches one of them towards nightfall, sees the flicker of the hearth-flame through the chinks in the walls; and at night, if the wind rises, he hears the roof of boughs shake to and fro in the midst of the great forest trees. Who would not suppose that this poor hut is the asylum of rudeness and ignorance? Yet no sort of comparison can be drawn between the pioneer and the dwelling which shelters him...he is, in short, a highly civilized being, who consents, for a time, to inhabit the backwoods, and who penetrates into the wilds of the New World with the Bible, an axe, and a file of newspapers. (de Tocqueville, 326)

In *Utopias: The American Experience*, authors Moment and Kraushaar recorded the rash of utopian communities that sprang up on the rapidly receding American frontier. Carving civilization from the wilderness was already leading to murmurs of disapproval over where this all was headed.

For despite its vastness, the wilderness was not in fact inexhaustible. As early as 1832, American adventurer and artist George Catlin warned that the immense herds of buffalo and the prairie stretching all the way from Mexico to Canada would both soon disappear without some kind of formal protection. He called for a "nation's park" forty years before Yellowstone was established as the country's first national park.

By the 1850s there arose a muted cry from artists, poets, and writers in response to the dystopian image resulting from the Industrial Revolution. The burgeoning muscle of technology, which reduced the toil required to conquer wilderness to a mere pittance of its romantic history, began to be a real force in society. Thoreau, Emerson, and others identified the American personality as interwoven into a fabric of pristine, untarnished, romantic Nature. They saw the vanishing primeval state of America as a real crime, and their message began to boil down to a sense of impending doom. "Not yet subdued to man, its presence refreshes him . . . in wildness is the preservation of the world," Thoreau wrote in 1851.

Perry Miller noted that growing movement. "Students of history of art recognize a doctrine that had, by 1847, become conventional among landscape painters in Europe, England, and America; that of the fundamental opposition of Nature to civilization, with the assumption that all virtue, repose and dignity are on the side of 'Nature' spelled with a capital and referred to as feminine—against the ugliness, squalor, and confusion of civilization, for which the pronoun was simply 'it.'"

By the 1870s preservationists such as John Muir implored politicians to create national parks and monuments in order to at least preserve some vestige of the dwindling grandeur of the West. Wrote Muir: "In God's wildness lies the hope of the world—the great fresh unblighted, unredeemed wilderness (Alaska Fragment, 1890).

Moment and Kraushaar captured the movement thusly: "The conservationists were in the end able to persuade the American people that preservation of the primeval forests and monumental landscapes was the proper concern of the federal government, in the interest of Americanism. . . . In this development, the wilderness, from being a desert of death and devils, has, with a variegated flora and fauna, living in peace, been converted into a kind of paradise. The devil now prowls outside the wilderness, the combination of selfish economic powers that would despoil the 'continental garden' for private greed."

Famed American writer Edward Abbey picked up that feeling in his book *The Monkey Wrench Gang* (Lippincott, 1975), which celebrated a team of ecoterrorists bent on destroying Glen Canyon Dam to allow the Colorado River to resume its natural flow at the bottom of the Grand Canyon. His somewhat self-styled character Doc Sarvis observed, "The wilderness once offered men a plausible way of life. . . . Now it functions as a psychiatric refuge." In Abbey's view, "We are caught in the treads of a technological juggernaut. A mindless machine, with a breeder reactor for a heart. . . . A planetary industrialism, growing like a cancer. Growth for the sake of growth. Power for the sake of power."

BACK TO THE LAND

It was probably inevitable that Americans would eventually head back to the land, as the noble idea of homesteading has never really left us. As long as 2,000 years ago, Roman citizens repeatedly fled the turmoil of the capital during the empire's more tumultuous times and headed out to the countryside.

In the late 1960s and early 1970s, a wave of urban counterculture refugees bailed on their lives in the city and took up alternative lifestyles in rural areas, not just in California but across North America and in Europe. In his book *Topophilia*, which translates to "love of place," author Yi-Fu Tuan applied principles of humanistic geography to describe the "back to the land" movement popular among the counterculture of the 1970s. Many of those farms, communes, and retreats were long ago disbanded or sold off, but the spirit of that small exodus persists.

Today, a small core of committed homesteaders has birthed a complete industry around self-sufficiency, permaculture, traditional farming, heirloom gardening, and preparing to survive various doomsday scenarios. From asteroid impacts to a potential zombie apocalypse, there is a strong sense that faster and faster semiconductors aren't the only answer to society's problems, and that it could all come crumbling down. Under those varied scenarios, when the inevitable "Sh*t Hits The Fan (SHTF)," the primitive skills of our ancestors will spring immediately back into play.

The worldwide "ecovillage" movement emphasizes sustainable practices in both rural and urban settings. Considerable research from multiple agricultural colleges and universities now exists to help struggling homesteads, and teams of researchers continue to make breakthroughs and establish new approaches to agriculture, energy, waste management, structural design, and other skills. Often, those advances are based on the primitive skills our ancestors practiced, and much of the time, practical geology is involved.

The immediate challenge is that changing weather patterns, longer and more persistent droughts, and increasingly catastrophic forest fires all stand in the way of long-term success off the grid. Just as energetic homesteaders push farther from the cities into deserts and mountains, the challenges there are increasing. It's no wonder that we are not only pushing into new environments on Earth but, at the same time, leaving the planet and broadening our ability to homestead on another planet.

THE MARS ROVER—SUPER ROCKHOUND

On February 18, 2021, the Perseverance rover landed at Jezero Crater on Mars with specific geologic duties. Its goal is to seek signs of ancient life, collect rock samples, and hopefully return those samples to Earth.

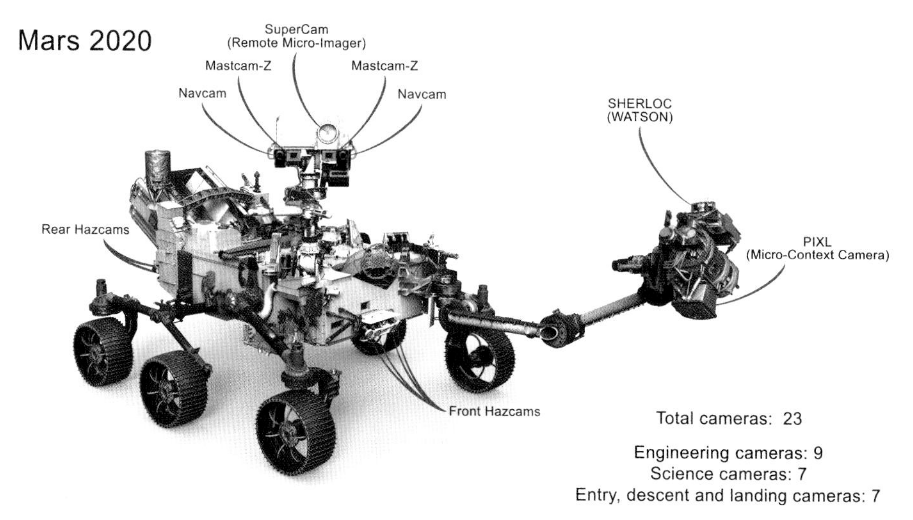

Mars Perseverance rover, showing the location of twenty-three cameras
NASA/JPL-CALTECH (IN THE PUBLIC DOMAIN)

In essence, it is an epic recon mission of practical geology, with rockhounding and homesteading in mind.

The rover carries seven primary payload instruments, twenty-three cameras, and two microphones, plus the Ingenuity mini-helicopter. Perseverance also carries a proof-of-concept experiment called Mars Oxygen In-Situ Resource Utilization Experiment (MOXIE) that will attempt to produce oxygen from carbon dioxide. The Mars Environmental Dynamics Analyzer (MEDA) instrument suite will gather data about weather, climate, and surface ultraviolet radiation and dust. Perseverance also packs an ultraviolet spectrometer, lasers, sensors, ground-penetrating radar, and a drill at the end of its robotic arm. The first homestead on Mars will indeed be data-driven.

COLONIZING THE MOON AND MARS

In the book and movie *The Martian* (self-published by Andy Weir, 2011; 20th Century Fox, 2015), astronaut Mark Watney (played by actor Matt Damon) finds himself marooned on Mars in the not-too-distant future. An epic sandstorm left him dazed and low on oxygen—his communications and life monitors knocked offline—and forced the rest of the crew to abandon him and their base.

Handy, resourceful, and stubborn, Watney exhibits all the positive attributes of an extraplanetary homesteader. He soon unleashes his formidable botany powers and converts Martian soil, human waste, and manufactured water into a healthy greenhouse. His attitude toward each problem is telling. He boldly vows to "Science the sh*t out of it" and solve one challenge at a time. He unearths a decades-old NASA probe named Pathfinder, revives it, and "phones home" using its older but still reliable technology, using handwritten messages, then more complex hexadecimal and ASCII code, and finally a software patch to link his vehicle to Earth.

When the book project was under way, author Andy Weir in part used "crowd-sourcing" with scientists worldwide to ensure the scientific accuracy of his premise and Watney's progression from doomed exile to hitchhiker thumbing a ride home when the crew circles back. At its heart it's an epic tale of human perseverance, but it's also a metaphor for homesteading on the frontier. Solve one problem at a time, carefully think through the science, keep your inventory of resources up to date, and keep experimenting.

In an October 2021 article at SciWorthy.com, Hunter Dulay described how scientists proposed a workaround for the lack of limestone on Mars to create Martian concrete from scratch. They proposed that colonists could donate blood plasma (and then collect the sticky protein serum albumin) and mix that with Martian rock dust called regolith to create concrete. They found a mix of 35 percent serum albumin and 65 percent regolith was superstrong, and would theoretically work with a 3D printer. Since the average healthy adult produces 12 to 25 grams (0.42 to 0.88 ounce) of plasma a day and can donate plasma as often as every few days, the team may have a clever workaround for "blood concrete." Further experimentation revealed that treating the serum albumin with urea (collected from urine) made the concrete even stronger.

FINAL THOUGHTS

The natural world is full of wonder, mystery, and beauty, but nothing currently known is indestructible, capable of turning lead into gold, able to aid in time travel, or bestow immortality. Comic book writers, fiction authors, and movie script creators have no such limits, and they have created an impressive list of invented materials. Their additions to the elemental table greatly improve on what we have at hand.

You don't have to look far on the internet to find a complete and growing list of all the fictional chemical elements, materials, isotopes, or subatomic particles that writers throughout history have created. Some of these elements are purely tongue in cheek, such as administratium, which is a statement about the invisible force that makes bureaucracy strong in the science community.

Balognium is a science fiction element invented to rescue a wavering plot. Dilithium, the driving force behind warp engines in *Star Trek*, gets a mention, as does jerktonium, which causes a bad attitude in *Sponge-Bob* cartoons.

Kryptonite, flubber, mythril, scrith, silverstone, Stygian iron . . . the list is as endless as the human imagination. Adamantium is the metal that powers Captain America's shield and Wolverine's claws. Vibranium is the elemental power behind Black Panther and the Kingdom of Wakanda. Transformium powers the Transformers; promethium is impervious to fire.

You get the idea—the periodic table might be great, but maybe there are some "what if?" elements out there that surely could improve things. In aerospace, there is a metal somewhere that is theoretically impossible but always sought: wish-alloy. Pilots sometimes joked that a new flight technology was made from unobtanium-reinforced wish-alloy.

But rather than yearn for rocks and minerals that don't exist, it's better to remain practical and work with the materials at hand. In this book, we've taken a pretty good ride. We've moved from crude hammerstones and pyrite fire starters to the epic constructions of the ancient world. We've advanced from preventing cholera to exploring the cosmos. Our rockhounding skills have progressed from eying a local gravel bar for arrowhead material to drilling holes on a distant planet thanks to the Mars Perseverance rover. And we've done it all using science, ingenuity, and sometimes a quick whiff of armpit juice. With luck, you'll never need 99 percent of the hacks and shortcuts listed here. But that obscure 1 percent could save your life.

REFERENCES

Alt, David, Donald W. Hyndman, et al. *Roadside Geology* [series]. Mountain Press [multiple books for different states and regions].

Andrews, John Joseph. *Finding Gold in California*. John Joseph Andrews publisher, 1980.

Bahn, Paul G. (ed). *100 Great Archaeological Discoveries*. Barnes & Noble, 1995.

Basque, Garnet. *Gold Panner's Manual*. Mr. Paperback, 1980.

Baumann, Stephen (ed.). *Geology—Super Review*. Research & Education Association, 2008.

Beci, Bharti. *Rocks and Minerals*. Penguin Random House, 2016.

Bryson, Bill. *A Short History of Nearly Everything*. Broadway Books, 2003.

Brzys, Karen. *Agates Inside Out*. Gitche Gumee Agate and History Museum, 2010.

DeVoto, Bernard. *The Journals of Lewis and Clark*. Houghton Mifflin, 1953.

Diamond, Jared. *Guns, Germs, and Steel: The Fates of Human Societies*. W.W. Norton, 1999.

———. *The World Until Yesterday: What We Can Learn from Traditional Societies*. Penguin Books, 2012.

Douglas, Jack. *Gold in Placer: How to Find It, How to Get It*. The Rosicrucian Press, Ltd., 1948.

Eckert, Allan W. *Earth Treasures* [series]. Harper & Row.

English, George Letchworth. *Getting Acquainted with Minerals*. McGraw-Hill Book Co., 1934.

Ettinger, L. J. *The Rockhound and Prospector's Bible: A Reference and Study Guide to Rocks, Minerals, Gemstones, and Prospecting*. L. J. Ettinger, 1992.

Fritzen, D. K. *The Rock-Hunter's Field Manual*. Harper and Row, 1959.

Harari, Yuval Noah. *Sapiens: A Brief History of Humankind*. HarperCollins, 2015.

Harker, Alfred. *Metamorphism: A Study of the Transformation of Rocks*. Chapman and Hall, 1974.

Gigantés, Philippe. *Power and Greed: A Short History of the World*. Carroll & Graf Publishers, 2002.

Johnson, Robert Neil. *Gold Diggers Atlas*. Cy Johnson & Son, 1971.

Kaplan, Robert D. *The Revenge of Geography: What the Map Tells Us About Coming Conflicts and the Battle Against Fate*. Random House, 2012.

Kassel, David. *A Slice Through America: A Geological Atlas*. Princeton Architectural Press, 2021.

Kaysing, Bill. *Great Hot Springs of the West*, 2nd edition. Capra Press, 1990.

King, Elbert A. *Space Geology: An Introduction*. John Wiley and Sons, Inc., 1976.

Kolbert, Elizabeth. *The Sixth Extinction: An Unnatural History*. Henry Holt, 2004.

Koschmann, A. H., and M. H. Bergendahl. *Principal Gold-Producing Districts of the United States*. US Geological Survey Professional Paper 610. US Government Printing Office, 1968.

Krajick, Kevin. *Barren Lands: An Epic Search for Diamonds in the North American Arctic*. Henry Holt and Company, 2002.

Kurlansky, Mark. *Salt: A World History*. Penguin Books, 2002.

MacPherson, John & Geri. *Naked into the Wilderness: Primitive Wilderness Living and Survival Skills*. Prairie Wolf Publishing, 1993.

Macwelch, Tim. *How to Survive Off the Grid: From Backyard Homesteads to Bunkers (And Everything in Between)*. WeldonOwen, 2019.

McManners, Hugh. *The Complete Wilderness Training Manual*, 2nd edition. Dorling Kindersley, 1994.

McPhee, John. *Annals of the Former World*. Farrar, Straus & Giroux, 1998.

Mitchell, James. *The Rockhound's Handbook*. Gem Guides Book Co., 2008.

Moclock, Leslie, and Jacob Selander. *Rocks, Minerals and Geology of the Pacific Northwest*. Timber Press, 2021.

Murray, Charles. *Human Accomplishment: The Pursuit of Excellence in the Arts and Sciences, 800 B.C. to 1950*. HarperCollins, 2004.

Nesbitt, Paul, Alonzo Pond, and William Allen. *The Survival Book*. Funk & Wagnalls, 1968.

Norton, O. Richard, and Lawrence Chitwood. *Field Guide to Meteors and Meteorites*. Springer-Verlag, 2008.

Notkin, Geoffrey. *Meteorite Hunting: How to Find Treasure From Space*. Aerolite Minerals LLC, 2011.

Osborne, Roger. *The Floating Egg: Episodes in the Making of Geology*. Jonathan Cape, 1998.

Pellant, Chris. *Rocks and Minerals*. Smithsonian Handbooks, Dorling Kindersley, Inc., 2002.

Pough, Frederick H. *Rocks and Minerals*. Houghton Mifflin Co., 1991.

Ridge, John D. *Ore Deposits in the United States: 1933–1967*. The American Institute of Mining, Metallurgical, and Petroleum Engineers, Inc., 1968.

Romaine, Garret. *Rocks, Gems, and Minerals of the Rocky Mountains*. FalconGuides, 2014.

———. *Rocks, Gems, and Minerals*, 2nd edition. FalconGuides, 2015.

———. *The Modern Rockhounding and Prospecting Handbook*, 2nd edition. FalconGuides, 2018.

———. *Basic Rockhounding and Prospecting*. FalconGuides, 2019.

———. *Geology Lab for Kids*. Quarto Publishing, 2017.

Sinkankas, John. *Gemstones of North America*. Von Nostrand Reinhold, 1959.

Strong, Emory. *Stone Age on the Columbia River*. Metropolitan Press, 1959.

Tomecek, Steve. *Everything Rocks and Minerals*. National Geographic, 2011.

US Geological Survey. *Mineral Resources of the United States—Calendar Year 1907*. Government Printing Office, 1908.

Williams, David B. *Stories in Stone: Travels Through Urban Geology*. Walker Publishing Co., 2009.

Winchester, Simon. *The Map that Changed the World: William Smith and the Birth of Modern Geology*. Harper Perennial; reprint ed., 2009.

INDEX

Italicized page numbers indicate illustrations.

ABOUT THE AUTHOR

Garret Romaine, an avid rockhound, fossil collector, and gem hunter, is the author of numerous rockhounding guides, including *Rockhounding Idaho*, *Basic Rockhounding and Prospecting*, *Gem Trails of Northern California*, and *Gem Trails of Oregon*. He has also written three gold panning titles: *Gold Panning California*, *Gold Panning Colorado*, and *Gold Panning the Pacific Northwest*.

Garret was a columnist for *Gold Prospectors* magazine for fifteen years and is the former executive director of the Rice Northwest Museum of Rocks and Minerals in Hillsboro, Oregon. He formerly served on the board of directors for the fossil-hunting organization North America Research Group. He holds a bachelor's degree in geology from the University of Oregon, a master's degree in geography from the University of Washington, and an MBA from Portland State University, where he still teaches technical writing.